Experimental Investigations on Particle Number Emissions from GDI Engines

Zur Erlangung des akademischen Grades
Doktor der Ingenieurwissenschaften

der Fakultät für Maschinenbau
des Karlsruher Instituts für Technologie (KIT)

genehmigte
Dissertation
von

Dipl.-Ing. Markus Bertsch
aus Karlsruhe

Tag der mündlichen Prüfung: 09.09.2016
Hauptreferent: Prof. Dr. sc. techn. Thomas Koch
Korreferent: Prof. Dr. rer. nat. Rainer Suntz

Forschungsberichte aus dem
Institut für Kolbenmaschinen
Karlsruher Institut für Technologie
Hrsg.: Prof. Dr. sc. techn. Thomas Koch

Bibliografische Information der Deutschen Nationalbibliothek

Die Deutsche Nationalbibliothek verzeichnet diese Publikation in der
Deutschen Nationalbibliografie; detaillierte bibliografische Daten sind
im Internet über http://dnb.d-nb.de abrufbar.

ISBN 978-3-8325-4403-4
ISSN 1615-2980

Logos Verlag Berlin GmbH
Comeniushof, Gubener Str. 47,
10243 Berlin
Tel.: +49 030 42 85 10 90
Fax: +49 030 42 85 10 92
INTERNET: http://www.logos-verlag.de

Vorwort des Herausgebers

Die Komplexität des verbrennungsmotorischen Antriebes ist seit über 100 Jahren Antrieb für kontinuierliche Aktivitäten im Bereich der Grundlagenforschung sowie der anwendungsorientierten Entwicklung. Die Kombination eines instationären, thermodynamischen Prozesses mit einem chemisch reaktiven und hochturbulenten Gemisch, welches in intensiver Wechselwirkung mit einer Mehrphasenströmung steht, stellt den technologisch anspruchsvollsten Anwendungsfall dar. Gleichzeitig ist das Produkt des Verbrennungsmotors aufgrund seiner vielseitigen Einsetzbarkeit und zahlreicher Produktvorteile für sehr viele Anwendungen annähernd konkurrenzlos. Nun steht der Verbrennungsmotor insbesondere aufgrund der Abgasemissionen im Blickpunkt des öffentlichen Interesses. Vor diesem Hintergrund ist eine weitere und kontinuierliche Verbesserung der Produkteigenschaften des Verbrennungsmotors unabdingbar.

Am Institut für Kolbenmaschinen am Karlsruher Institut für Technologie wird deshalb intensiv an der Weiterentwicklung des Verbrennungsmotors geforscht. Übergeordnetes Ziel dieser Forschungsaktivitäten ist die Konzentration auf drei Entwicklungsschwerpunkte. Zum einen ist die weitere Reduzierung der Emissionen des Verbrennungsmotors, die bereits im Verlauf der letzten beiden Dekaden um circa zwei Größenordnungen reduziert werden konnten aufzuführen. Zum zweiten ist die langfristige Umstellung der Kraftstoffe auf eine nachhaltige Basis Ziel der verbrennungsmotorischen Forschungsaktivitäten. Diese Aktivitäten fokussieren gleichzeitig auf eine weitere Wirkungsgradsteigerung des Verbrennungsmotors. Der dritte Entwicklungsschwerpunkt zielt auf eine Systemverbesserung. Motivation ist beispielsweise eine Kostenreduzierung, Systemvereinfachung oder Robustheitssteigerung von technischen Lösungen. Bei den meisten Fragestellungen wird aus dem Dreiklang aus Grundlagenexperiment, Prüfstandversuch und Simulation eine technische Lösung erarbeitet.

Die Arbeit an diesen Entwicklungsschwerpunkten bestimmt die Forschungs- und Entwicklungsaktivitäten des Instituts. Hierbei ist eine gesunde Mischung aus grundlagenorientierter Forschung und anwendungsorientierter Entwicklungsarbeit der Schlüssel für ein erfolgreiches Wirken. In nationalen als auch internationalen Vorhaben sind wir bestrebt, einen wissenschaftlich wertvollen Beitrag zur erfolgreichen Weiterentwicklung des Verbrennungsmotors beizusteuern. Sowohl Industriekooperationen als auch öffentlich geförderte Forschungsaktivitäten sind hierbei die Grundlage guter universitärer Forschung.

Zur Diskussion der erarbeiteten Ergebnisse und Erkenntnisse dient diese Schriftenreihe, in der die Dissertationen des Instituts für Kolbenmaschinen verfasst sind. In dieser Sammlung sind somit die wesentlichen Ausarbeitungen des Instituts niedergeschrieben. Natürlich werden darüber hinaus auch Publikationen auf Konferenzen und in Fachzeitschriften veröffentlicht. Präsenz in der Fachwelt erarbeiten wir uns zudem durch die Einreichung von Erfindungsmeldungen und dem damit verknüpften Streben nach Patenten. Diese Aktivitäten sind jedoch erst das Resultat von vorgelagerter und erfolgreicher Grundlagenforschung.

Jeder Doktorand am Institut beschäftigt sich mit Fragestellungen von ausgeprägter gesellschaftlicher Relevanz. Insbesondere Nachhaltigkeit und Umweltschutz als Triebfedern des ingenieurwissenschaftlichen Handelns sind die Motivation unserer Aktivität. Gleichzeitig kann er nach Beendigung seiner Promotion mit einer sehr guten Ausbildung in der Industrie oder Forschungslandschaft wichtige Beiträge leisten.

Im vorliegenden Band 02/2016 berichtet Herr Bertsch über experimentelle Untersuchungen zur Reduktion von Partikelanzahlemissionen beim Ottomotor mit Direkteinspritzung. Messungen an einem Einzylinder-Forschungsaggregat mit kombiniertem Einsatz von Partikelanzahlmessgerät, Partikelgrößenverteilungsmessung sowie optischer Diagnostik und thermodynamischer Analyse ermöglichen dabei die detaillierte Analyse der Partikelbildung und -oxidation. Hierzu werden zahlreiche optische Diagnosetechniken zur Visualisierung der Gemischbildung (Mie-Streulicht, High-Speed PIV) sowie der Rußbildung und -oxidation (High-Speed Imaging, Lichtleitermesstechnik) eingesetzt.

Zunächst werden zwei Injektoren mit unterschiedlichem hydraulischen Durchfluss und identischem Spraytargeting in einer Einspritzdruckkammer charakterisiert und bewertet. Im Fokus der experimentellen Arbeiten am Versuchsmotor steht dann der Betrieb bei erhöhter Motorlast und geringer Motordrehzahl. Hierbei stellen die geringen Strömungsgeschwindigkeiten im Brennraum, bedingt durch die geringe Motordrehzahl, sowie die große eingebrachte Kraftstoffmasse eine zentrale Herausforderung für den Gemischbildungsprozess dar. Einen wesentlichen Teil der Arbeit stellt deshalb die detaillierte Analyse des Gemischbildungsprozesses, welcher als Summe aus Kraftstoffeinbringung, Interaktion der Ladungsbewegung mit dem eingebrachten Kraftstoff sowie der Kraftstoffbeschaffenheit beschrieben werden kann, dar.

Maßnahmen zur Optimierung der Gemischbildung und Minimierung der Partikelemissionen werden abgeleitet und bewertet. Neben der gezielten Beeinflussung der Ladungsbewegung durch das Aufprägen einer gerichteten Strömung und der Variation der Ventilöffnungszeitpunkte und -öffnungsverläufe, wird auch die Einspritzung gezielt beeinflusst. Hierzu wird neben einer Reduktion des hydraulischen Durchflusses des Injektors auch eine Erhöhung des Einspritzdruckes auf bis zu 500 bar diskutiert.

Die Untersuchungen zeigen am Beispiel des aufgeladenen Betriebs grundlegende Auswirkungen und Potenziale verschiedener Variationsparameter bezüglich der innermotorischen Emissionsreduzierung auf. Durch die gleichzeitige Analyse der Ladunsgbewegung und eine thermodynamische Analyse können die Ergebnisse auf weitere Motoren übertragen werden.

Karlsruhe, im Dezember 2016   Prof. Dr. sc. techn. Thomas Koch

Vorwort des Autors

Die vorliegende Arbeit entstand während meiner Tätigkeit als wissenschaftlicher Mitarbeiter am Institut für Kolbenmaschinen des Karlsruher Instituts für Technologie. Sie ging aus dem von 2013 bis 2015 bearbeiteten und von der Forschungsvereinigung Verbrennungskraftmaschinen finanzierten Vorhaben "Partikel bei Otto DI II" hervor.

Mein besonderer Dank gilt an erster Stelle Herrn Prof. Dr. sc. techn. Thomas Koch für das mir entgegengebrachte Vertrauen und die mir gebotene Freiheit bei der Gestaltung und Durchführung der Arbeit. Außerdem hat er mich ermuntert, den Blick auch einmal über den Mikrokosmos des eigenen Forschungsgebietes hinaus wandern zu lassen, wofür ich ebenfalls danken möchte [16]. Für die Übernahme des Korreferats und das große Interesse an meiner Arbeit danke ich Herrn Prof. Dr. rer. nat. Rainer Suntz. Für seinen unermüdlichen Einsatz als Obmann des FVV-Projektes und die vielen fachlichen Diskussionen danke ich herzlich Herrn Dr.-Ing. Daniel Sabathil. Des Weiteren möchte ich mich bei meinem Gruppenleiter Herrn Dr.-Ing. Amin Velji bedanken. Die Dienstreisen nach Paris und Kyoto werden mir in guter Erinnerung bleiben.

Mein Dank gilt allen Kollegen, die zum Gelingen dieser Arbeit, sowie zur angenehmen Arbeitsatmosphäre am Institut beigetragen haben. Besonders erwähnt seien hier natürlich auch die Aktivitäten außerhalb des Institutsalltages, wie die Radelrunde, gelegentliche Konzertbesuche oder die Zuberabende. Speziell die geselligen Abende trugen erheblich zur Steigerung des Betriebsklimas bei.

Persönlich bedanken möchte ich mich an dieser Stelle bei meinem langjährigen Büromitbewohner Dr.-Ing. Kai W. Beck. Die vielen Gespräche rund um den Verbrennungsmotor und weit darüber hinaus haben den Alltag ungemein bereichert. Unvergessen sind mir auch die Reisen rund um die Welt, bei der wir unsere Sonderforschungsprojekte der Fachwelt vorstellen durften [11, 12, 13, 17, 19]. Mit Dr.-Ing. Clemens Hampe kann man auch super forschen [19, 73, 74] und feiern sowieso. Dr.-Ing. Stefan Berlenz fährt mit seinem Mountainbike schneller den Berg hoch als ich mit meinem Rennrad und organisiert die besten Ausflüge in den Westen Frankreichs. Sowohl Dr.-Ing. Florian Schumann als auch Dr.-Ing. Helge Dageförde haben einen sehr guten Musikgeschmack. Mit zweit genanntem kann man die Musik sogar im tiefsten Schlamm genießen. Außerdem danke ich den beiden für die Korrektur dieses Werkes sowie die zahlreichen Diskussionen rund um Partikelemissionen. Christian Disch weiß sehr viel über Optik und mit ihm kommt man fast auf den Mt. Fuji. Mit Dr.-Ing. Philipp Hügel und Kai Scheiber kann man bei einem ordentlichen Espresso super diskutieren. Dr.-Ing. Gerald Banzhaf hingegen trinkt zwar keinen Kaffee, mit ihm zu diskutieren hat aber auch immer Spaß gemacht. Ivica Kraljevic danke ich für den inter-institutionellen Austausch, der mich um so manche Stunde Schlaf gebracht hat. Mit Florian Sobek und Johannes Dörn-höfer wurde so manche Stunde in spannende Bastelprojekte investiert. Meinem Freun-deskreis aus Uni-Zeiten danke ich für die Organisation der besten Winterurlaube und für all die schönen Freizeitaktivitäten. Hannah McBeath und Ruth Petters-Raskob danke ich zusätzlich für den sprachlichen Feinschliff der Arbeit.

Helge Rosenthal weiß bei sämtlichen technischen Problemen neben einem guten Rat auch noch, wo er anpacken muss. Eduard Oberländer operiert mit ruhiger Hand abgerissene optische Zugänge aus Versuchsmotoren. Für die schnelle und präzise Fertigung zahlreicher kleiner und großer Versuchsteile danke ich der gesamten Ferti-gungsabteilung um Ernst Hummel. Ohne die Hilfe von Werner Kruggel und Christoph Schramm wäre die Abgasmesstechnik den Belastungen meiner zahlreichen Messreihen nicht gewachsen gewesen. Für die nicht weniger zahlreichen Umbau- und Reparatur-maßnahmen am Prüfstand sowie die immer angenehme Zusammenarbeit gilt mein Dank Markus Weber und Christian Stahl.

Im Laufe meiner Jahre am Institut durfte ich mit einer Vielzahl an Studenten zusam-men arbeiten, deren Auflistung den Rahmen dieses Vorwortes übersteigt. Ein paar Studenten, die über längere Zeit wesentlich zum gelingen dieser Arbeit beigetragen haben, möchte ich dennoch gesondert danken. So konnte ich jahrelang auf die sehr gute Zuarbeit von Matthias Helmich bauen. Sowohl beim Schrauben am Prüfstand als auch bei der Datenverarbeitung wusste er sich nützlich zu machen. Fabian Rauber half mir das Aggregat für höhere Lasten zu ertüchtigen. Philipp Werner hat die Grundlagen für ein Simulationsmodell gelegt, dass tatsächlich funktioniert hat. Markus Kollmer kam zwar meist erst zur Mittagszeit, war mir dann aber eine große Hilfe.

Ein nicht in Worte zu fassender Dank gebührt meinen Eltern Franz und Susanne Bertsch. Ihre Liebe und bedingungslose Unterstützung in allen Lebenslagen haben meine Entwicklung erst ermöglicht. Ich liebe euch! Außerdem haben sie es geschafft, dass wir Geschwister, Rebecca, David, Klaus und ich, blindes Vertrauen zueinander haben. Auch meinen Geschwistern, meinen Neffen Frederik und Oskar sowie meiner Nichte Tessa gilt es an dieser Stelle ein Dank auszusprechen. Nicht nur für ihre Un-terstützung und die motivierenden Worte, sondern auch dafür, dass sie mir immer gezeigt haben, was wirklich wichtig ist. Abschließend bedanken möchte ich mich bei meiner Freundin Liv Wittkopf, die mich stets zum lachen bringt und auch in den Endzügen dieser Arbeit nicht die Geduld mit mir verloren hat - Worte vermögen nicht zu beschreiben, was du mir bedeutest.

Karlsruhe, im Dezember 2016   Markus Bertsch

Contents

Nomenclature

Physical Quantities

Symbol	Unit	Description
cfr	kHz	Camera frame rate
CO	%	Carbon monoxide concentration
CO_2	%	Carbon dioxide concentration
CoV IMEP	%	Coefficient of Variation of IMEP
d	mm	Diameter
D_{char}	mm	Nozzle hole diameter
Δp	MPa	Pressure difference
E_{AKE}	m^2/s^2	Average kinetic energy
E_{TKE}	m^2/s^2	Turbulent kinetic energy
EOI	°CA aTDCf	End of Injection
evc	°CA aTDCf	Exhaust Valve Closing timing
evo	°CA aTDCf	Exhaust Valve Open timing
η_a	kg/(m s)	Dynamic viscosity gas
η_f	kg/(m s)	Dynamic viscosity fuel
FDA	°CA	Flame Development Angle (Ignition timing to MFB05%)
IMEP	MPa	Indicated Mean Effective Pressure
ivc	°CA aTDCf	Intake Valve Closing timing
ivo	°CA aTDCf	Intake Valve Open timing
iws	pixel x pixel	Interrogation window size
λ		Air-fuel equivalence ratio
MFBxx%	°CA aTDCf	Mass Fraction Burned xx%
n		Number of vectors
NO_x	ppm	Nitrogen oxides concentration
O_2	%	Oxygen concentration
Oh		Ohnesorge number
p_{Sys}	MPa	Injection pressure
PN_{CPC}	$\#/cm^3$	Particle number concentration measured with an AVL 489 APC Advanced
PN_{EEPS}	$\#/cm^3$	Particle number concentration measured with a TSI 3090 EEPS
$\dot{Q}_{stat}$	l/min	Hydraulic flow
RBA	°CA	Rapid Burn Angle (MFB05% to MFB90%)
Re		Reynolds number

Symbol	Unit	Description
ρ_a	kg/m^3	Density gas
ρ_f	kg/m^3	Density fuel
rpm	$1/min$	Revolutions per minute
σ	N/m	Surface tension
SOI	$°CA_{aTDCf}$	Start of Injection
Std. of Int.	Counts	Standard deviation of Intensity
THC	ppm	Hydrocarbon concentration
ti	ms	injection duration
v_{avg}	m/s	Average velocity
v_i	m/s	Velocity of vector i
v_{rel}	m/s	Relative velocity
$We_{gas/Fl}$		Weber number of gas / fuel

Abbreviations and indices

Abbreviation	Description
AiF	Arbeitsgemeinschaft industrieller Forschungsvereinigungen "Otto von Guericke" e.V.
BDC	Bottom Dead Centre
BMEP	Brake mean effective pressure
BSFC	Brake specific fuel consumption
CCD	Camera with a CCD Chip (charge-coupled device)
CMOS	Camera with a CMOS Chip (Complementary metal-oxide-semiconductor)
CPC	Condensation Particle Counter
CVS	Constant Volume Sampling
DBE	Double bond equivalent
DoE	Design of Experiments
DPF	Diesel particulate filter
EEPS	Engine Exhaust Particle Sizer
EGR	Exhaust gas recirculation
FVV	Forschungsvereinigung Verbrennungskraftmaschinen e.V.
GDI	Direct Injection Gasoline Engine
HFI	High Frequency Ignition
HiL	Hardware in the loop
HP-EGR	High Pressure EGR
IFKM	Institute of Internal Combustion Engines - Karlsruhe
IMEP	Indicated mean effective pressure
IWC	Institute of Hydrochemistry - Munich
JRC	Joint Research Center
KIT	Karlsruhe Institute of Technology

Abbreviation	Description
LIF	Laser Induced Fluorescence
LII	Laser Induced Incandescence
LP-EGR	Low Pressure EGR
LVK	Institute of Internal Combustion Engines - Munich
NEDC	New European Driving Cycle
PAH	Polycyclic aromatic hydrocarbons
PEMS	Portable Emission Measurement System
PFDS	Partial Flow Dilution System
PFI	Port Fuel Injection
PIV	Particle Image Velocimetry
PM	Particulate Mass
$PM_{0.1}$	Particle mass with a diameter < 0.1 μm
$PM_{2.5}$	Particle mass with a diameter < 2.5 μm
PM_{10}	Particle mass with a diameter < 10 μm
PMP	Particle Measurement Programme
PN	Particle Number
RDE	Real Driving Emissions
REGR	Reformate exhaust gas recirculation
RON	Research Octane Number
SMD	Sauter Mean Diameter
SMPS	Scanning Mobility Particle Sizer
SI	Spark ignition
TCI	Transistor Coil Ignition
TDC	Top Dead Centre
TKE	Turbulent kinetic energy
TPA	Three Pressure Analysis
UDC	Unified Driving Cycle
UNECE	United Nations Economy Comission for Europe
V.P	Vapor pressure
VPR	Volatile Particle Remover
WLTP	Worldwide Harmonized Light-Duty Vehicles Test Procedure
WOT	Wide Open Throttle

1 Introduction

Air pollutants can be categorised in primary and secondary air pollutants. The primary pollutants are directly emitted to the atmosphere, whereas secondary pollutants are formed in the atmosphere from precursor gases. Main precursor gases are ozone (O_3), secondary NO_2 and secondary PM, which are generated from SO_2, NO_x, NH_3 and volatile organic compounds (VOC) [48]. The European Environment Agency states that "Primary PM originates from both natural and anthropogenic sources. Natural sources include sea salt, naturally suspended dust, pollen and volcanic ash. Anthropogenic sources, which are predominant in urban areas, include fuel combustion in thermal power generation, incineration, domestic heating for households and fuel combustion for vehicles, as well as vehicle (tyre and brake) and road wear and other types of anthropogenic dust. Black carbon (BC) is one of the constituents of fine PM and has a warming effect. BC is a product of incomplete combustion of organic carbon as emitted from traffic, fossil fuels and biomass burning and industry. "[48]

The 2015 report by the European Environment Agency about "Air quality in Europe" shows a decrease of both PM_{10} and $PM_{2.5}$ from 2004 to 2013, as shown in Figure 1.1. The main emitters of air pollutants in Europe are transport, energy production and distribution, industry, agriculture, households and waste. The development of PM emissions since 2004 for these sectors are shown in Figure 1.2 relative to the emission level of 2004. The relative emission of PM_{10} are shown in the upper graph, whereas the emission of $PM_{2.5}$ are shown in the lower graph.

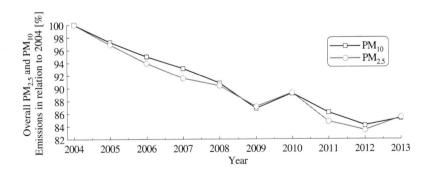

Figure 1.1: Development of PM_{10} and $PM_{2.5}$ in relation to 2004 in Europe [48]

Of these sectors, industry and transport have strongly reduced both emissions of $PM_{2.5}$ and PM_{10}. However, the transport sector further accounts for 13 % of PM_{10} and 15 %

of $PM_{2.5}$ emissions respectively. The "Traffic and air quality contribution of traffic to urban air quality in European cities" revealed that about 50 % of PM_{10} emission and about 22 % of $PM_{2.5}$[1] emission of the traffic sector were caused by non-exhaust gas emissions [70]. Studies showed that even with zero tail-pipe emissions, traffic will contribute to PM emissions significantly [34, 114]. A study by Rexeis and Hausberger predicts a proportion of about 90 % of total PM will be caused by non-exhaust sources by 2020 [157]. In addition, emissions from shipping within Europe may contribute about 15 % of the $PM_{2.5}$ [47].

The commercial, institutional and households sector dominates the emission of primary $PM_{2.5}$ and PM_{10}. In contrast to the transport sector, the PM emissions from this sector increased since 2004, probably caused by higher use of wood and other biomass combustion for heating in some countries [48]. The emission of PM, both of the agriculture and the energy production sector, increased since 2004.

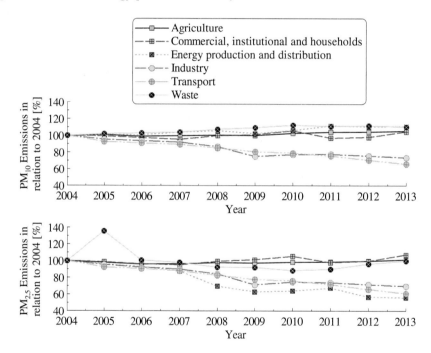

Figure 1.2: Development of PM_{10} and $PM_{2.5}$ for the different sectors in relation to 2004 in Europe [48]

It should be noted that not only the amount of pollutants emitted but also the proximity of the source, emission conditions[2], dispersion conditions, topography and other

---

[1]The number of cities providing $PM_{2.5}$ measurements was only about 10. However, the authors state that the results of $PM_{2.5}$ did not differ substantially from PM_{10}

[2]for example height and temperature

factors influence the contribution of the emitter to the ambient air concentrations [48]. Thus emission sectors such as traffic and households with a low emission height make a larger contribution to ambient air concentrations of the pollutants.

Concerning the transport sector, PM emissions were driven mostly by Diesel engines without DPF in the past. For gasoline engines with external mixture formation, studies showed that the PM emissions are generally on a low level [67, 68, 72, 160]. Compared to Diesel engines, nanoparticle emissions from gasoline engines are much more speed and load dependent and emit a higher amount of smaller particles even though the particle mass is on a lower level. As downsizing and downspeeding are the most promising solutions to reduce the fuel consumption of modern gasoline engines [163], the engines operate more frequently at high engine loads and low engine speeds compared to engines with a larger displacement. For the higher engine load, a higher amount of fuel needs to be injected. In combination with the low in-cylinder charge motion caused by the low engine speeds, the mixture preparation is challenging. The parameters influencing the particle number emission at high engine loads are summarised in Figure 1.3.

The mixture formation process in gasoline engines is a result of the fuel injection, evaporation and homogenisation with the surrounding charge air flow. In homogeneous gasoline engines with direct injection (GDI), the inflammation process starts the regular combustion process after the mixture formation. All these processes can be influenced by changing parameters of the engine hardware, engine settings or operating parameters.

In this work, the influence of the following parameters on the mixture formation, inflammation, combustion and emissions formation were investigated:

To influence the mixture formation process, different *inlays to generate a large scale charge motion* were adapted to the engine. Additionally, the charge exchange was influenced by the *valve timings* and the *valve open time*. The turbulence level at injection timing was influenced by changing the *maximum intake valve lift* and in addition the valve timings were changed by using the phase shifters. To increase the amount of *residual gas*, the exhaust gas back-pressure at high engine load was varied.

As gasoline engines are prone to knocking at high engine load, several techniques to enhance the knocking resistance are available. One promising solution to reduce the in-cylinder temperatures that was investigated is the use of *recirculated exhaust gas*. Thereby the caloric of the charge is changed and thus the temperature at the end of compression is reduced. In state-of-the-art engines, a *fuel enrichment* for high engine loads is used to cool the in-cylinder gas and to keep the engine from damage. Thereby the increased injection duration and higher fuel mass are assumed to increase the inhomogeneity of the air-fuel mixture and the lack of oxygen is assumed to increase the soot formation.

For the enrichment at high engine load, the fuel mass needs to be increased to *reduce the air-fuel ratio (λ)*, thus influencing the fuel injection process due to longer injection duration and possible wall impingement. To reduce a possible wall impingement, either the *injection timing* can be optimised for each injection duration, or the *injection pressure* can be increased to shorten the injection duration for the same amount of fuel injected. The injector characteristics, such as the nozzle hole number and diameters as well as the targeting of the nozzle holes also influences the injection process. Due to the higher injection pressure available in recent times[3], a reduction of the injector static flow is possible to enhance the spray break-up and thus increase the mixture formation process.

Both the *physical and chemical properties* of the fuel and the *fuel conditions* (temperature and pressure) significantly influence the injection process and the evaporation of the fuel. Furthermore, the boundary conditions such as the *temperature of oil and cooling* influence the in-cylinder temperatures, and the evaporation of liquid fuel at operation with wall impingement. Besides the mixture formation, the inflammation of the air-fuel mixture shows significant impact on the combustion process and thus on the emission formation. To enhance the *inflammation process*, a high-frequency ignition system (Corona ignition) was compared to a conventional transistor coil ignition.

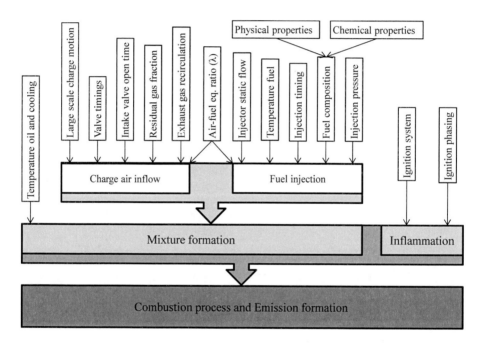

Figure 1.3: Sources of particle emissions in GDI engines

---

[3]Euro 5 GDI engines mostly operated with an injection pressure of 20 MPa, whereas the next generation GDI systems are capable of increasing the injection pressure up to 35 MPa

2 Fundamentals on particle emissions from GDI engines

After the basics of soot formation and particle emissions, this section gives a short summary of the particle measurement programme and the actual measurement devices are discussed. Thereafter the focus is on the particle formation processes in GDI engines and the influencing parameters, especially the mixture formation process and the inflammation. Exhaust gas recirculation is discussed in general and a brief overview of the effects on the particle formation discussed in literature is given. Finally, the influence of the fuel composition on the particle formation process is discussed.

2.1 Combustion generated particles

The particles emitted by combustion engines consist typically of three distinct types separated by their diameter, labelled 'nucleation mode', 'accumulation mode' and 'coarse mode'. Coarse mode particles are of varying nature, such as rust from the exhaust system. They are not emitted directly but formed from the other two modes. The predecessors are stored within the exhaust system, become attached to each other and re-enter the exhaust flow as larger particles. Because of this storage-release process, the coarse mode is an inconsistent emission. However, it is suggested that they consist of a solid core and an outer layer of volatile material. As the coarse mode is of unsettled state, artificial nature and comparative rarity, these particles have been little studied [42].

Historically, nucleation mode particles have been less studied, because they are at the limit of detection for many instruments. The consistency of these particles is yet a topic of research. A literature survey by Giechaskiel, Manfredi and Martini gives an overview of the current state of research on nucleation mode particles [60, 61]. They state:

"The structure of primary particles is sometimes different (more amorphous) and unburned hydrocarbons or volatile organics can be found. This means that differences in the thermal pre-treatment (temperature, residence time of PN systems) might lead to different results. A lot of studies have found a solid core mode with older and modern Diesel engines, both at low and high loads. Solid core mode is often observed at gasoline engines with port fuel injection (G-PFI) and it is assumed to originate from the metals of the lube oil or from fuel additives. At GDIs a shoulder at 10 nm to

20 nm appears quite often. For mopeds very often the size distribution after thermal pre-treatment peaks at or below 20 nm. It should be mentioned that in many studies it was recognised that the 'solid' core mode was re-nucleation artefact of the PMP method and the dilution factors employed."

Accumulation mode particles consist of a collection of much smaller 'primary' particles[1]. The typical size of the primary accumulation mode particles ranges from about 20 nm to 50 nm. It should be mentioned that accumulation mode particles vary in size because they contain greater or fewer numbers of primary particles, not because the primary particles vary in size. The number of primary particles to form an agglomerate is not a fixed number and varies from tens through hundreds to thousands. The morphology of agglomerates is also of a diverse nature, while the surface is coated by a layer of volatile or semi-volatile material.

The chemical composition of particles can be defined according to the four-layered conceptual model by Eastwood [42], depicted in Figure 2.1.

The five main components of the particles are presented subsequently according to

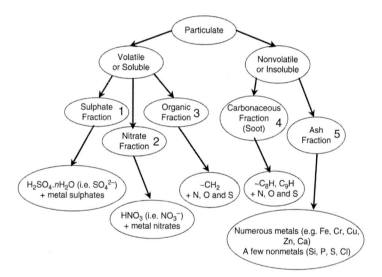

Figure 2.1: Conceptual composition of particle composition according to Eastwood [42]

the discussion by Eastwood [42]. As the sulphate bound in the fuel mainly contributes to the sulphate fraction, it is important to mention that this model was based on Diesel particles.

1 SULPHATE FRACTION

The sulphate fraction mainly consists of water-soluble sulphates, or the SO_4^{2-} ion, with

---

[1] Also referred as 'spherules'; even though they are not exactly spherical, they quite closely approximate sphericity [42]

the main component sulphuric acid, H_2SO_4. There is a correlation of the amount of sulphuric acid and the amount of water on the filter, thus dependant on the humidity in the immediate environment of the filter. Therefore, prior to gravimetric measurements, the filters must be conditioned for a certain period of time in a closely defined environment.

2 NITRATE FRACTION

The nitrate fraction consists of water-soluble nitrates, like the NO_3^- ion, with the main compound nitric acid, HNO_3. Compared to the sulphate fraction, the amount of the nitrate fraction is usually lower. Additionally, nitric acid has a lower boiling point compared to sulphuric acid, showing a greater volatility.

3 ORGANIC FRACTION

Based on the assay method, the organic fraction is also known as soluble organic fraction (SOF)[2], or volatile organic fraction (VOF)[3]. However, the mass of SOF is usually quite close to the mass of VOF, even though the separation processes are based on different properties that do not necessarily generate the same final result. The organic fraction consists of several hundred compounds, such as alkenes, alkanes, alcohols, esters, ketones, acids and aromatics. Even lighter C_4- to C_8- compounds were detected, which should be in the gaseous state, supporting the assumption that surface interactions are strong.

4 CARBONACEOUS FRACTION

The title 'carbonaceous' implies that this fraction is predominantly, but not exclusively carbon. In the research community, this fraction is also titled as 'soot', 'graphite carbon', 'elemental carbon' or 'black carbon'. Although the absence of any coherent carbon compounds is meant. Therefore, in this work, 'soot' and 'carbonaceous fraction' are synonyms. More details on the composition, formation and oxidation of the carbonaceous fraction are given in section 2.2.

5 ASH FRACTION

The ash fraction covers inorganic compounds or elements, such as metals and a few non-metals and is a mixture of highly variable composition. A method to characterise the ash-fraction is to burn all the particles and thereafter analysing the remains, which is the 'incombustible ash'. The ash fraction is mostly caused by additives of fuel and oil as well as the elemental composition of the working fluids[4] itself.

The ratio for each of these fractions on the total particle emission depends on the composition of the working fluids, the engine operating parameters, the wear performance of the engine as well as influences of the exhaust gas after treatment devices.
Besides the chemical composition, particles get classified according to their diameters.

---

[2]If assayed by dissolution in an organic solvent
[3]If assayed by heating or volatilizing
[4]Fuel and oil

As smaller particles do not contribute strongly to the particle mass, the legislative saw the requirement to limit the particle number and thus the emission of fine particles. Because of their small diameters, these particles are respirable and represent a considerable health risk. In Figure 2.2 a typical distribution of the particle emissions of a Diesel engine is given by Kittelson et al. [103]. Thereby particle size distributions are plotted in three ways: according to number, area and mass. The particles can be divided into three modes: nuclei ('nucleation mode'), accumulation and coarse mode as already mentioned at the beginning of the chapter. Fundamental is the domination of different ranges by number and mass. Concerning particle numbers, most of the particles reside in the nucleation mode, while for the particle mass, the maximum is found in the accumulation mode. In general, the particles are classified by their diameter as follows:

- PM_{10} for particles with a diameter below $10\,\mu m$

- $PM_{2.5}$ for particles with a diameter below $2.5\,\mu m$ ('fine particles')

- $PM_{0.1}$ for particles with a diameter below $100\,nm$ ('ultra-fine particles')

- Particles with a diameter below $50\,nm$ are known as 'nanoparticles'

Even though the term 'diameter' implicates particles to be of spherical shape, particles are more likely to show branched or reticular structures. Therefore, the measured diameter of particles by the actual measurement systems is a comparative value and gives the reference diameter of an ideal round particle, which would behave comparable to the measured particle. According to Hinds, different definitions for the reference diameter are common [80]:

- HYDRODYNAMIC DIAMETER - diameter of a reference particle showing equal diffusion properties

- AERODYNAMIC DIAMETER - diameter of a reference particle with the density of $1\,\frac{g}{cm^3}$ and equal descent rate

- ELECTRICAL MOBILITY DIAMETER - diameter of a reference particle with a well defined charge that shows equal mobility in an electrical field

The different diameters can be converted to each other as exemplary shown by Jimenez et al. [90] and McMurray et al. [132]. However, the existence of different definitions with different values for the diameter supports the assumption that shape, mass and surface of particles with equal diameter can show significant differences. Furthermore, especially the nucleation mode particles can consist of a large number of volatile particles.

As the nucleation mode particles are mostly dominated by volatile particles, the sampling position and the thermal pre-treatment of the exhaust gas can influence the measured size distributions.

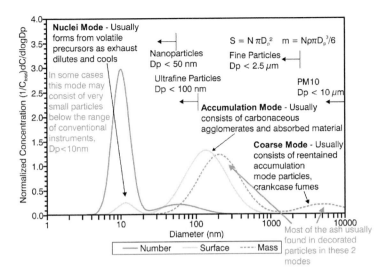

Figure 2.2: Typical engine exhaust particle size distribution by mass, number and surface area by Kittelson et al. [103], figure adapted by Dageförde [30]

2.2 Soot formation

Of the above-mentioned particle fractions, ash and soot are generated within the combustion process of engines, while the other components form later on in the exhaust system or even not before entering the surrounding air [42].

Even though the formation process of soot is up until now not determined in every detail, it is known that polycyclic aromatic hydrocarbons (PAH) play a major role in the process [20]. The formation process of PAH is dominated by acetylene (C_2H_2), which is present in a larger amount in fuel-rich flames. By reactions with CH and CH_2, C_3H_3 is formed, which tends to form a benzene ring (aromatic ring) by recombination and rearrangement. By abstraction of hydrogen and acetylene addition (HACA mechanism) [51], these PAH molecules grow planar. Spatial structures with a graphite-like structure are formed from the plain PAH molecules and grow further by the HACA mechanism. These build up three-dimensional structures are known as primary soot particles. Due to surface growth and coagulation, which means a conglomeration of particles, the soot formation continues and extends. The growth of singular particles is limited, even though the reasons are not fully discovered yet. This process is followed by the agglomeration of colliding particles, building up loosely structured agglomerates. The mechanism is shown schematically in Figure 2.3.

The soot output is thereby dependant on carbon to oxygen ratio and temperature, as shown in Figure 2.3 on the right hand side by Böhm et al. [22]. A maximum output of soot is thereby obvious for rich air-fuel equivalence ratios and temperatures from 1600 K to 1700 K. For lower temperatures, the formation of radical precursors like C_3H_3 is inhibited. For higher temperatures, these precursors are pyrolysed and oxidised. Therefore, the soot formation is limited to temperatures from about 1000 K to 2000 K [197].

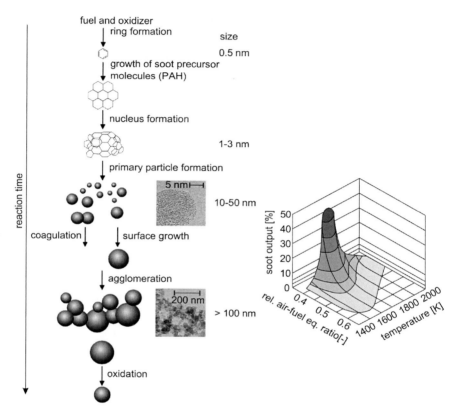

Figure 2.3: Soot formation process according to Bockhorn [20] with pictures of soot structures created with HR-TEM (left figure) and soot output dependency of temperature and air-fuel ratio by Böhm et al. [22] (right figure)

Soot oxidation can occur at the precursor, nuclei and particle stages of the soot formation process. Heywood states, that in general the rate of heterogeneous reactions such as the oxidation of soot depends on the diffusion of reactants to and products from the surface as well as the kinetics of the reaction [79]. There are many species in or near the flame that could oxidise soot, such as O_2, O, OH, CO_2 and H_2O. For the premixed, stoichiometric combustion in gasoline engines, the soot oxidation process is assumed to be dominated by OH, as the concentration of O and O_2 is on a low level [79]. However, in compression ignition engines a large fraction of the soot formed is oxidised within the cylinder due to the excess oxygen [181]. For the dependency of the soot oxidation process from oxygen partial pressure and temperature, a number of models are available [42, 99]. As it has been proven difficult to follow the oxidation of soot aerosols in flames, studies of bulk samples of pyrographite can be used to understand the soot oxidation process. The semi-empirical formula of Nagle and Strickland-Constable [140] has been shown to correlate pyrographite oxidation for oxygen partial pressures below 0.1 MPa and temperatures in the range of 1100 to 2500 K, and is therefore often used to describe soot oxidation mechanisms [79].

2.3 Particle number measurement

Historically, particulate matter emissions were defined as a mass collected on a filter. As the limits for the particulate mass have become more stringent, it is very difficult to achieve repeatable results. Leach et al. gave a good overview of the measurement uncertainties [116]:

"The EU particulate matter limit (4.5 mg/km) corresponds to about 250 µg (allowing for dilution) on a filter that might weigh 100 mg. A balance is needed with a readability of 0.1 µg, and corrections even need to be made for buoyancy. The filters have to be weighed in a room with temperature and humidity control; even so, the permitted variation in humidity (from 37 % - 53 %) can lead to a change in mass of 7 µg; [58]. The number and size are also significant since the deposition efficiency in the respiratory system is greatest for small particles. So it is potentially the smaller particles that are the most harmful, so this is another reason for the introduction of legislation that limits the total number of particles that are emitted [42]."

By the end of the 20th century, a lot of investigations were done concerning the particle emissions of gasoline engines [67, 68, 71, 72, 126, 127, 128, 160]. These studies have shown that compared to Diesel engines, nanoparticle emissions from SI engines are much more speed and load dependant and emit a higher amount of smaller particles. Thereby particulate matter emissions (PM) are on a very low level. All the cars taken into the study emitted less than 2 mg/km PM, which was well below the limit of 50 mg/km PM at that time. The mean particle size for all vehicles tested in the study ranged from 35 nm to 65 nm by number and 100 nm to 300 nm by mass respectively. Peak particle emissions were found correlating to short periods of heavy acceleration. These peaks coincided with peaks in carbon monoxide and nitrogen oxide emissions. Kittelson et al. published measurements of an on-road and laboratory investigation in 2006 [104]. They used seven cars of model years from 1984 to 1999 and engine displacements from 1.8 l to 5.0 l. Kittelson et al. stated that it was impossible to measure a significant particle signature above background for all cars under highway conditions. However, during full throttle accelerations, the tested cars emitted size distributions comparable to heavy duty Diesel engines. The cars emitted lower particle number concentrations in the upper end of the accumulation mode, where most of the particle mass is found. Although, older cars with higher mileage emitted a greater number of nuclei mode particles compared to the lower mileage cars. Once again, Kittelson et al. showed that no simple relation between particulate mass and particle number emissions exists. Looking at cold start conditions, the vehicles tested emitted 5 to 30 times the particulate mass emissions compared to hot (ambient) conditions. The authors conclude, that cold-start conditions must be considered in a test cycle for low emitting gasoline-powered engines. Comparing on-road and chassis dynamometer tests, the authors found the emissions on particle volume (mass) base by the on-road test to be between cold and hot chassis dynamometer tests. For particle number emissions, on-road CPC[5] and SMPS[6] measurements were found to be higher compared to the chassis dynamometer. The authors assume that the nucleation mode particles are underestimated using traditional test methods.

As consequence of the studies concerning particle emissions, several governments decided to limit particle number emissions for combustion engines. In Europe, the JRC guided a research group to define both the measurement technique and the emission limits for passenger cars in the European Union for the exhaust gas legislation Euro 6 [58]. It was shown in this inter-laboratory exercise, that the particulate mass measurement had an unsatisfactory repeatability ($\approx 55\,\%$). Furthermore, even using high efficiency filters in the dilution tunnel, the background levels were found similar to the emissions of the vehicles ($\approx 0.4\,\mathrm{mg/km}$). Thereby most of the mass collected on the filters was volatile and less than $10\,\%$ was soot. A new method measuring particle number was therefore approved, focusing on non-volatile particles using a volatile particle remover (VPR)[7] and a CPC with $50\,\%$ cut-off efficiency at 23 nm ($90\,\%$ at 40 nm). This new method showed good intra-lab and inter-lab reproducibility of $\approx 40\,\%$ and $\approx 25\,\%$ respectively. These values are similar to other gaseous emissions like carbon monoxide and hydrocarbon. Even though no difference in the repeatability variabilities was found comparing the reference particle measurement system and the labs own systems, a $15\,\%$ difference in the mean number emission levels of the vehicle was found. The authors state, that the calibration procedure for the particle number measurement systems should be better defined in order to ensure the measurement of the absolute value of the vehicle emissions. Based on these findings, the Economic Commission for Europe of the United Nations (UNECE) published the *Uniform provisions concerning the approval of vehicles with regard to the emission of pollutants according to engine fuel requirements* [92, 192]. Based on this legislative in 2014, the new emissions legislation Euro 6b was introduced and for the first time, a limit was set for particle number emissions of $6 \cdot 10^{11}\,\#/\mathrm{km}$.

The roadmap of the European legislative concerning PN emissions is shown in Figure 2.4. Besides the measurement of the PN emissions in the new European driving Cycle (NEDC), currently a monitoring of real driving emissions (RDE) is ongoing.

---

[5]Condensation Particle Counter
[6]Scanning Mobility Particle Sizer
[7]hot dilution $> 10\!:\!1$ at $150\,^\circ$C and thermal pretreatment at $300\,^\circ$C to $400\,^\circ$C

These RDE measurements are needed because laboratory tests do not accurately reflect the amount of air pollution emitted during real driving conditions. The compliance factors for RDE were announced by the European Commission in October 2015 [46]. In a first step, a conformity factor of 2.1 for new models by September 2017[8] is fixed as well as the second step with a factor of 1.5 for all new models by January 2020[9]. Although these factors are comparatively high, the European Commission states that the "agreement by Member States on the allowed divergence between the regulatory limit measured in real driving conditions and measured in laboratory conditions is still a significant reduction compared to the current discrepancy (400 % on average)" [46]. Additionally, the NEDC will be replaced by the Worldwide Harmonised Light-Duty Vehicles Test Procedure (WLTP) including the Worldwide Harmonised Light-Duty Vehicles Test Cycle (WLTC) [191].

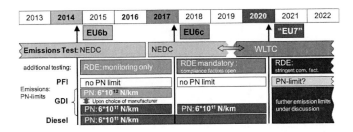

Figure 2.4: Roadmap of the European legislative concerning PN emissions [203]

In 2012, Giechaskiel et al. published a review on the measurement of automotive non-volatile particle number emissions within the European framework [59]. The measurement setup as well as the particle transformation along the measuring path is shown in a simplified diagram in Figure 2.5. Starting from the combustion in the cylinder of the engine to the measurement instrument is connected to the full-flow dilution tunnel (Constant Volume Sampling, CVS) for the determination of the PN and PM emissions. In the combustion chamber, primary particles (spherules) with a diameter of about 10 nm to 30 nm form via the pyrolysis of fuel and lubricant. As the exhaust gas temperature in the exhaust tailpipe and the transfer tube to the full-dilution tunnel is still high, the primary particles usually coagulate and form agglomerates. The exhaust gas is mixed with dilution air in the CVS. Thereby volatile material can absorb or condense on the agglomerates and / or nucleate to form a distinct volatile nucleation mode. Therefore, at the end of the dilution tunnel, a tri-modal particle size distribution can be measured: The nucleation mode, mainly consisting of volatile droplets, the accumulation mode, mainly consisting of carbonaceous agglomerates with condensed or absorbed hydrocarbon, and the coarse mode, which consists of reentrained

---

[8]for new vehicles by September 2019

[9]for new vehicles by January 2021

deposited particles or wear material. While the accumulation mode particles account for a large part of the mass, the nucleation mode particles can dominate the number concentration under certain conditions [102].

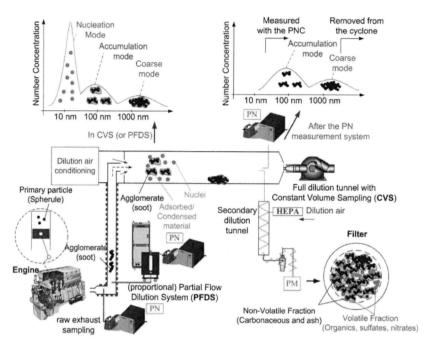

Figure 2.5: Typical sequence of particle transformation from the engine to the measurement location [59]

However, for research purposes on engine test benches, a full flow dilution tunnel is not always available. Thus a common approach is to measure the raw exhaust gas at the tailpipe. Giechaskiel et al. showed that this procedure achieves particle number measurement results close to the legislated procedure using a CVS or partial flow dilution system (PFDS) [57]. However, special attention needs to be given to the sampling position and the proper conditioning of the chemically aggressive exhaust aerosol.

Even though this PN measurement method has been proven robust, Giechaskiel et al. state that several issues need further consideration. First of all, the volatile particles have been proven to be responsible for some adverse health effects. These particles are not measured with the current PN measurement setup and need well-defined conditions to be measured, such as sampling from tailpipe with constant-dilution conditions. Therefore, the authors state that more research needs to be done until systems ready for legislative use will be available. The second issue is the measurement of non-volatile particles with a diameter below 23 nm. Simply removing the cut-off size of the CPC would be advisable only if ensured that all volatiles have been removed and thus will not falsely be measured as non-volatiles. However, lowering the cut-off size is challenging not only due to possible volatile artefacts, but also from the perspective of calibration and losses of the CPCs and VPRs.

As there were unanswered questions about the PMP measurement procedure, the Forschungsvereinigung Verbrennungskraftmaschinen e.V. (FVV) initiated the project *"Investigations and evaluation of different particle number measurement systems"* in 2011. Within the framework of this research project, studies on the measurement principle of the particle number concentration in Diesel exhaust gas were conducted, following and investigating the European regulations 83 and 49. The project was divided into two parts: investigations on model aerosols (Institute of Hydrochemistry Munich, IWC) and on real Diesel exhaust gas (Institute of Internal Combustion Engines Munich, LVK) [122, 123, 124, 125]. Investigations on the model aerosol by IWC demonstrated a 15 % difference in counting efficiency of PMP-conform particle counters from different manufacturers. These results correspond well to findings of Mr. Dierks presented at the ETH Conference on Combustion Generated Nanoparticles in Zürich in 2013 [37].

In the FVV project, using different volatile particles, a difference of about 50 % could be observed in the response characteristics of the CPC. By implementing a catalytic stripper and an evapouration tube, a total separation of the volatile particles from the aerosol flow was possible. However, VPRs of different manufacturers did not achieve comparable results. At the engine test bench of LVK, a heavy duty engine was equipped with four different PN-measurement systems. Systematic differences in the total count of about 20 % were as well identified as dynamic and size-dependant influences on the systems. In conclusion, LVK and IWC proposed a revised calibration guideline for the CPCs and a more precise definition of the working parameters for the VPRs to reduce the measurement uncertainty. Furthermore, besides an increased frequency of the single component calibrations [91], a calibration of the entire system should be considered.

Concerning the measurement for the RDE tests, new measurement devices need to be developed. These new measurement devices are known as Portable Emission Measurement System (PEMS). For the gaseous exhaust gas components, the systems are recently well developed [130], while PN-PEMS is still in validation process [159, 195]. Current PN-PEMS systems show a variation of about ± 60 % with respect to a PMP compliant system, as shown by Vlachos et al. [195]. Even the PN measurement principle for the RDE certification is not yet decided. Right now, two systems compete for PN-PEMS: A condensation particle counter (CPC) [35] and a diffusion charger [196].

2.4 Sources of PN emissions in GDI engines - Scientific knowledge

In the following section, the current state of knowledge concerning particle emission from GDI engines is shown. Firstly the mechanisms of particle formation in GDI engines are discussed, followed by basics on mixture formation and inflammation in GDI engines. After the discussion of the basics of exhaust gas recirculation (EGR), publications concerning the influence of EGR on PN emissions are presented. Finally the PN-Index developed by Aikawa et al. [2] is presented, which describes the relation of the fuel composition to PN emissions.

2.4.1 Mechanism of PN formation

Soot formation generally depends on the local temperature and on the local air-fuel ratio [20]. Theoretically fuel-rich zones should not exist for the homogeneous, stoichiometric operation of a GDI engine. According to earlier investigations, it is known that spray-wall interaction is one of the main causes for soot formation in GDI engines [89, 110, 113, 193].

Spray-wall interaction becomes critical concerning soot formation only if the time before ignition is too short to evapourate and sufficiently homogenise the liquid fuel. During the following combustion process most of the injected fuel is oxidised. After the regular combustion, the temperature in the combustion chamber is on a high level. In combination with the low oxygen content after combustion, due to the stoichiometric operation, ideal conditions for soot formation exist. Thus even small amounts of fuel stored in the wall films can lead to significantly increased particle number emissions. Contrary to Diesel engines, the soot can not be oxidised after the combustion because of the low oxygen content [112].

Wiese et al. published a work on the influencing factors on PN emissions [203]. A summary image of their findings is shown in Figure 2.6. Especially for cold engine operation, wall impingement on piston, intake valves and liner are critical. For hot engine operation, fuel impingement on the piston and the deposit formation at the injector tip is of great importance in terms of soot formation.

Jiao and Reitz made a numerical approach to study the effects of wall films on the piston surface caused by early injection timings and thus impingement of liquid fuel on the particle formation process in GDI engines [89]. As the flame propagates towards the piston, all the regions where the flame front passed become burnt regions, releasing free radicals at high temperature. These high temperature regions enhance the liquid film vapourisation and form fuel-rich zones. As the oxygen concentration in the burned areas is low, the fuel oxidation is limited, resulting in a diffusive combustion. Schulz et al. investigated the effects of gasoline wall films and spray-wall interaction

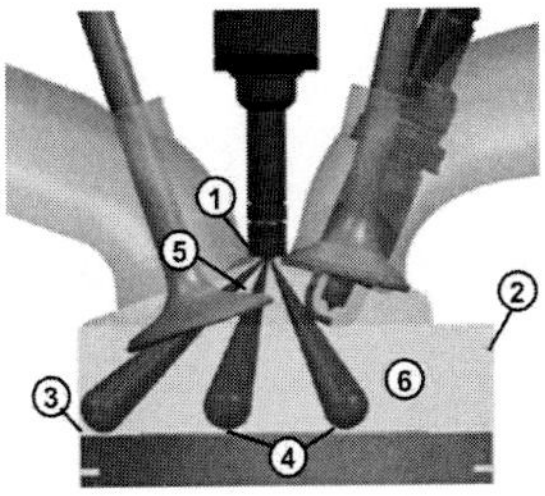

PN & HC raw emissions* homogneous mode multi hole injectors		PN		HC (raw)	
		cold	warm	cold	warm
1	injector	+	+++	+	0
2	liner	++	+	++	+
3	fireland	+	0	+++	++
4	piston	+++	++	++	0
5	intake valves	++	+	+	0
6	gas phases	0	0	+	0

*regular mode, not catalyst heating or engine start

Figure 2.6: Sources of PN and HC emissions in GDI engines for hot and cold engine conditions according to Wiese et al. [203]

on a hot plate using infra-red thermography [172]. Thereby they varied boundary conditions in a wide range and used Design of Experiment (DoE) methods to describe fundamental physical phenomena. Besides the initial temperature, the injection pressure, the distance from injector to the hot plate and the angle of the injector to the plate play important roles, regarding the wall film area and the wall film mass.

Another source for soot emissions in GDI engines are inhomogeneities in the gas phase caused by insufficient mixture preparation. These can be caused for example by an interaction of the spray and the open intake valve. Other reasons for inhomogeneities are for instance a too short period of time for mixture preparation because of a late injection timing, an insufficient level of charge motion in the cylinder, or a low injection pressure. These inhomogeneities usually show a stochastic character. Thereby single working cycles with high soot emissions increase the total particle number concentrations significantly [29].

Miyashita et al. developed a sampler for TEM analysis to obtain a sufficient number of soot particles even at low particle number levels (10^5 #/cm³) [138]. They measured particles of the exhaust gas of a single cylinder GDI engine for three injection timings, (normal, advanced and retarded) and quantitatively analysed their morphology and nanostructure. A comparison to the SMPS measured particle size distributions was also performed. They state:

"The distributions and their variation trend with the fuel injection timing of TEM-based aggregate gyration diameter qualitatively well corresponded to that of SMPS-measured particle size, except that numerous nanoparticles smaller than 23 nm, [...] and not regulated under EURO6, are observed only in the SMPS measurements. These nanoparticles likely correspond to volatile components which are expected to disappear during the vacuuming prior to the TEM observation. These qualitative correspondence between SMPS and TEM results ensure that the TEM analysis conducted in the present study is reliable."[138]

Concerning the different injection timings and their particle size distributions, Miyashita et al. suggest that GDI particles are formed at relatively low temperature and are exhausted without experiencing oxidation. Further they assume:

"The decrease of both aggregate and primary particle sizes depending on the fuel injection timing is therefore considered due to the difference in residence time for soot particles in the in-cylinder regions with temperatures suitable for soot formation and growth."

2.4.2 Mixture formation in GDI engines

In GDI engines, the liquid fuel is injected into the combustion chamber and therefore the mixture formation process depends on the injection system and its characteristics as well as on the interaction with the inflowing air. The in-cylinder charge motion thereby strongly affects the secondary break-up of the fuel and thus the evapouration process and the homogenisation of the mixture.

Fuel injection

For homogeneous operation in GDI engines, the fuel is injected early in the intake stroke. The injection and mixture formation are therefore temporally and spatially separated from the combustion. To accomplish a suitable mixture formation process resulting in a stable and low-emission combustion process, fine atomisation and fast evapouration of the fuel and at the same time a proper mixing of the fuel with the surrounding air is required. Besides the in-cylinder charge motion, the fuel injection process is therefore of great importance.

The research and development on the spray formation and break-up in combustion engines focussed mainly on Diesel engines in the past. However, as the mechanisms compare well with gasoline operation, it is reasonable to transfer the spray break-up of a Diesel spray, as shown in Figure 2.7, to a GDI injection system. The graphic by Baumgarten [8] shows the lower part of an injection nozzle with the needle, sac hole and injection hole. Immediately after leaving the nozzle hole, the fuel jet breaks-up into a conical spray and can be classified in primary and secondary break-up.

The primary break-up of high-pressure injection systems is dominated by the cavitation as well as turbulence and results in large ligaments and droplets that form a dense spray near the nozzle. Cavitation and turbulence are affected by injection pressure, the geometry of the nozzle and the physical properties of fuel and surrounding medium [43]. The secondary break-up is dominated by aerodynamic forces caused by the relative velocity between the droplets and the surrounding medium.

PRIMARY BREAK-UP

To describe these complex interdependencies, dimensionless parameters have been established. The liquid Weber number describes the dependency of the break-up of

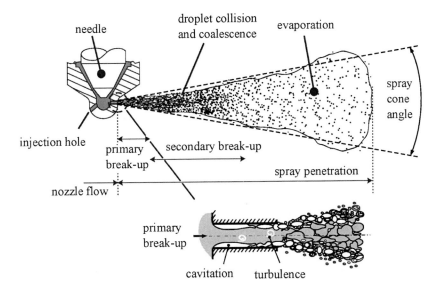

Figure 2.7: Break-up of a full-cone diesel spray by Baumgarten [8]

fluids with low viscosity, like gasoline, from the surface tension σ_{Fl} and the aerodynamic forces. The aerodynamic forces are described with the square of the relative velocity v_{rel}, the density of the liquid ρ_{Fl} and the nozzle hole diameter D_{char}.

$$We_{Fl} = \frac{v_{rel}^2 \cdot \rho_{Fl} \cdot D_{char}}{\sigma_{Fl}} \tag{2.1}$$

The Reynolds number, shown in equation 2.2, describes the flow-field of liquids in a tube flow by the relation of the inertial forces to the viscous forces and can be used to describe the flow-field in the nozzle holes (laminar, turbulent) [171]. The inertial forces are thereby calculated with the jet velocity v_{rel}, the density of the liquid ρ_{Fl} and the nozzle hole diameter D_{char}. The viscous forces correspond to the dynamic viscosity of the liquid η_{Fl}.

$$Re = \frac{v_{rel} \cdot \rho_{Fl} \cdot D_{char}}{\eta_{Fl}} \tag{2.2}$$

By the elimination of the jet velocity, Ohnesorge derived a dimensionless number which bears his name. It is calculated using the relevant properties such as the dynamic viscosity η_{Fl}, surface tension σ_{Fl}, density of the liquid ρ_{Fl} and nozzle hole diameter D_{char} [141]. It can also be expressed using the Reynolds number and the liquid Weber number, as shown in equation 2.3.

$$Oh = \frac{\eta_{Fl}}{\sqrt{\sigma_{Fl} \cdot \rho_{Fl} \cdot D_{char}}} = \frac{\sqrt{We_{Fl}}}{Re} \tag{2.3}$$

Ohnesorge used the parameters Re and Oh to describe the break-up mechanisms of a liquid jet in a diagram. By additionally using the liquid to gas density ratio on a third axis, Reitz derived the graph shown in Figure 2.8 [155, 156]. With increasing jet velocity, the Rayleigh, wind-induced and atomisation regime are passed. By the usage of the liquid to gas density ratio, the tendency for a faster break-up with higher gas density is obvious. However, the influence of the nozzle characteristics is not considered by using these theoretical parameters.

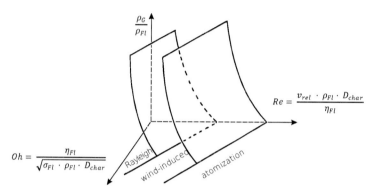

$$Re = \frac{v_{rel} \cdot \rho_{Fl} \cdot D_{char}}{\eta_{Fl}}$$

$$Oh = \frac{\eta_{Fl}}{\sqrt{\sigma_{Fl} \cdot \rho_{Fl} \cdot D_{char}}}$$

Figure 2.8: Schematic chart of influence of gas density on breakup regime boundaries by Reitz [155, 156]

For gasoline direct injection, the atomisation regime is of particular relevance. Theoretically, the atomisation regime is reached, when the intact surface length approaches zero. A conical spray develops and the intake core, or at least a dense core consisting of large liquid fragments may still be present several nozzle diameters downstream the nozzle [8]. For gasoline direct injection, an early break-up with a low penetration is required to realise a fast mixture formation and to reduce fuel impingement. Different publications showed, that for a low length-to-diameter ratio and high conicity, the spray angle is wider and the break-up enhanced [118, 171, 173, 202].

SECONDARY BREAK-UP

The secondary break-up is induced by the aerodynamic forces which act on the droplets due to the relative velocity between droplet and gas. The surface tension acts against the aerodynamic forces. The relation of aerodynamic forces and surface tension is described using the gas Weber number We_{gas} and differs to the liquid Weber number due to the usage of the gas density ρ_{gas} instead of density of the liquid and the droplet diameter $D_{droplet}$ instead of the nozzle hole diameter. It is known from experimental investigations, that different droplet break-up modes exist, depending on the Weber number [117, 146]. In engine sprays, all of these break-up mechanisms occur. However, most of the disintegration processes take place near the nozzle at high Weber numbers, while further downstream the Weber numbers are significantly smaller. This is caused by reduced droplet diameters due to evapouration and previous break-up, and because of a reduction of the relative velocity due to drag forces.

From start of injection on, a phase transition from liquid to gaseous phase of the fuel takes place. A higher surface-to-volume ratio thereby increases the evapouration process. As gasoline fuel is a multi-component fuel, the evapouration depends both on pressure and temperature and is mostly described with evapouration curves for constant pressure. With higher pressure, the evapouration curves shift to higher temperatures [173]. Pressure and temperature show opposite influence on the evapouration process.

Due to early or late injection timing, non-optimised spray targeting or insufficient time or space for evapouration, an impingement of liquid fuel on the combustion chamber walls can occur. The liquid fuel can evapourate fast, if the wall temperature is 40 °C to 50 °C above the evapouration temperature of the relevant fuel component [43]. Further increasing the temperature does not increase the evapouration process due to the Leidenfrost effect [8]. If the liquid fuel is not evapourated in time, due to lack of time or cold temperatures, the liquid fuel oxidises under lack of oxygen in a diffusive combustion and increases the soot formation.

Optimisation of the injection strategy [150], of nozzle type [31] and characteristics [202] and by an increase of the injection pressure [12, 24, 105, 129, 173], the primary break-up of the jet can be enhanced. Due to the subsequent increase of impulse of the fuel droplets, higher droplet velocities occur and therefore increase the aerodynamic forces. This leads to smaller droplets with higher surface-to-volume ratios, enhancing the evapouration process. Due to the increased relative velocity between droplet and surrounding gas, the partial pressure on the droplet surface is reduced, enhancing the convective material and heat exchange and thus further increases the evapouration process.

In-cylinder charge motion

The flow-field inside a combustion engine influences the fuel jet break-up and evapouration of the fuel, as shown in section 2.4.2, the heat transfer to the cylinder wall, the transport of the mixture and the inflammation and combustion process. In combustion engines, two types of flow can be separated: directed[10] and undirected flows. Separation of the directed flow from undirected flow (turbulence) is a matter of definition and is important because the speed of flame propagation is related to small-scale fluctuations.

Directed flows show a defined direction in the shape of a vortex structure, which is generated during the intake stroke. Depending on the position of the axis of rotation, the large scale charge motion is called tumble (rotational axis perpendicular to the cylinder axis) or swirl (rotational axis parallel to cylinder axis). Due to interactions with the surrounding flow field, the large vortex structures disintegrate into smaller

---

[10]in this work referred to as large scale charge motion

vortex structures and exchange kinetic energy [106]. The energy is thereby exchanged cascade-shaped from larger to smaller vortexes. The smallest vortex structures finally dissipate and exchange the turbulent kinetic energy into heat [109].

Using a tumble motion, the rotational velocity increases in the compression stroke due to the change of directions in BDC and the following reduction of vortex diameter caused by the reduced cylinder volume. During the compression stroke, the vortex structure gets compressed and changes direction following the geometry of the combustion chamber. Thereby large vortex structures dissipate into smaller vortex structures [43].

For the swirl motion, the radius of the vortex is not influenced by the compression. Therefore, the intensity of the swirl motion only dissipates during the compression due to interactions with other (undirected) flows inside the cylinder.

Disadvantage of the large scale charge motion is the reduction of the volumetric efficiency due to the reduced flow area [7].

Undirected flows are non-stationary, three dimensional vortex structures, characterised by high cycle-to-cycle variations. In combustion engines, the flow involves a complicated combination of turbulent shear layers, recirculating regions and boundary layers [79]. The inlet valve gap is the minimum area for the flow into the cylinder, so gas velocities in the valve gap are the highest velocities during the charge exchange process. The gas issues from the valve opening into the cylinder as a conical jet and the radial and axial velocities in the jet area are about ten times the mean piston speed [79].

Using direct injection systems with a high injection pressure, the fuel is injected with high velocities into the combustion chamber, inducing turbulent flow structures in the periphery of the fuel jet [39].

An additional cause of turbulent flows inside a combustion chamber is the squish flow, which is a radially inward or transverse gas motion towards the end of the compression stroke [79]. The squish flow is induced, as a portion of the piston and the cylinder head approach each other closely.

2.4.3 Inflammation in GDI engines

Transistor coil ignition (TCI)

The state-of-the-art system in gasoline engines to start the combustion process is the transistor coil ignition with a spark plug in the combustion chamber. In TCI systems, a spark can arc from one plug electrode to the other only if a sufficiently high voltage is applied. In this spark discharge, the plasma is heated up and expands thermally (high-temperature plasma), whereby locally conditions of auto ignition are reached [178, 179]. In the case of thermal plasma, there are sufficient exchanges of kinetic energy among electrons, ions and molecules such that their temperatures are in a thermal equilibrium state [176]. Therefore, approximately half of the energy induced is lost in the form of thermal loss to the electrodes as a result of the rise in gas temperature, due to the small initial flame kernel being located close to the electrodes

[79]. Following the initial plasma expansion, the first flame propagation is laminarly dominated and propagates comparatively slow [79].

Concerning the improvement of TCI systems, research has been done on increasing spark energy, applying multi-point or multi-pulse ignition and lengthening the spark discharge duration[176].

High-frequency ignition (HFI)

The high-frequency ignition system (HFI) consists of a resonant circuit in which the combustion chamber acts as a capacitor. To generate an electric field with changing polarities, an AC voltage in single-digit MHz range is applied. Compared with the TCI system, the tip of the igniters act as the centre electrode, whereas the combustion chamber acts as the ground electrode [169]. This tip-plate arrangement results in a strongly inhomogeneous electric field around the tip of the igniter, as shown in the upper part of Figure 2.9.

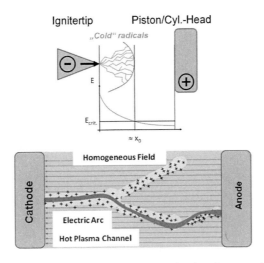

Figure 2.9: Schematic diagram of a corona discharge (top) and conventional spark discharge (bottom) [45], adapted by Wolf et al. [206]

Hampe et al. showed that the gas volume activated by the corona discharge is up to three orders of magnitude higher compared to the electric arc of a TCI system with a spark plug [75].

Hampe et al. describe the inflammation of HFI in comparison to TCI as follows [74]:

"The field strength approaches values up to 300 kV/mm near the tips of the electrode. The corona discharge is a special type of high-pressure discharge. Instead of using the thermal effect to oxidise the atoms and molecules in the combustion chamber and thus to initiate combustion [...], HFI ionises the particles by overcoming the binding energy of a valence electron and removing an electron [...]. When the concentration of free electrons is sufficient enough an avalanche effect takes place which entails a self-sustained gas discharge. This is known as the Townsend mechanism [161]. This electron impact dissociation reaction leads to a faster reaction rate and therefore a shorter burn delay [79, 176, 178]."

Compared to the plasma channel of a TCI system, the resulting non-thermal plasma has a high total resistance, resulting in a low amperage and thus hardly any heating of the plasma channels. For the corona discharge, only the almost inertia-less electrons sustain high kinetic energy and temperature, as described by Shiraishi [176] and shown in Figure 2.10.

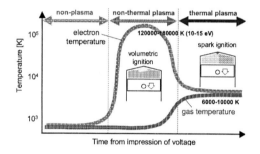

Figure 2.10: Schematic diagram of two kinds of plasma by Shiraishi et al. [176]

Since the density of radicals is not homogeneous, its gradients lead to locally different burning velocities. The resulting small-scale turbulence accelerates the inflammation from the beginning [93].
The advantages of HFI compared to TCI systems are shown in a number of publications focussing on lean burn concepts [25, 77, 148], homogeneous operation at high engine load [63, 206], enlargement of EGR dilution rate [206] and even in small two-stroke engines [73].

2.4.4 Tip-Sooting

Using multi-hole injectors, another possible reason for soot emissions is a deposit formation at the tip of the injector, known as 'tip-sooting'. The deposits are formed as a result of a liquid fuel film on the tip. Using a clean injector, the fuel evapourates almost completely before the flame front reaches the injector. However, if deposits are formed at the tip, fuel is stored within the porous structure of the deposits. The evapouration process is delayed and leads to an insufficient mixture preparation of the fuel stored. The fuel is oxidised after the initial combustion under oxygen deficiency and leads to increased soot formation. Measurements by Kinoshita et al. using a swirl injector described the mechanism in 1999 [101]. They describe the nozzle temperature as the major influence on deposit formation:

"When the nozzle temperature is lower than the 90 vol. % distillation temperature of the fuel, some fuel evapourates, and some of the fuel remains in a liquid state. Thus, the deposit precursors are maintained in a state of dispersion in the fuel. [...] On the other hand, when the nozzle temperature is higher than the 90 vol. % distillation temperature of the fuel, most of the fuel evapourates. The deposit precursors cohere to each other, and they adhere strongly on the nozzle hole wall." [101]

Measurements by Berndorfer et al. [10] and Wiese et al. [203] showed that this effect can increase the emitted particle number concentrations at steady state operation by more than one order of magnitude. Thereby neither the hydraulic flow nor the spray pattern are influenced by the deposits.

Piock et al. showed that an increased injection pressure is capable to reduce the deposit formation [147]. And even when deposits are build up at the injector tip, increased injection pressure up to 40 MPa showed a PN reduction.

Investigations concerning the deposit formation were also realised by Dageförde [33]. After running the engine about 500 minutes at 10 MPa injection pressure, the particle number emissions increased by about 360 %, while the PM emissions increased by about 510 %. After this test, the injector had deposits formed on the injector tip.

However, these processes are strongly influenced by the combustion chamber and injector geometry as well as the flow conditions close to the injector tip and the temperature profile. Due to these complex mechanisms of deposit formation and oxidation, tip-sooting is not discussed in the following work. In order to eliminate possible distortions by tip-sooting effects, the injector tip is cleaned prior to each test and reference operating points are regularly measured.

2.4.5 Exhaust gas recirculation

Basics of EGR

Exhaust gas recirculation (EGR) can be used in gasoline engines to control NO_x emissions and to reduce the fuel consumption. The EGR gets mixed with the fresh air before the cylinder and acts as diluent in the unburned gas mixture thus reducing the peak gas temperatures. Hence the peak temperature reached in the combustion chamber varies inversely with the EGR mass fraction.

In consequence, engine knocking can be reduced in gasoline engines by using EGR. A first study on this topic was published by Ricardo as early as 1919 with an experimental investigation on the effect of EGR on knock control [158]. Ricardo described how an inert gas added to the charge can "insulate the particles of fuel and air from each other, thereby delaying the rate of flame propagation", which made a significant increase in the compression ratio possible without the engine being more prone to knocking. Engine knocking is caused by autoignition in the end gas, which is heavily influenced by the end gas temperature and pressure-time histories, as Rothe [162] and Mittal [137] showed. End gas temperature and pressure-time history are influenced by three factors, as identified by Grandin [66]: heat transfer, compression and chemical reactions. As EGR increases the in-cylinder mass of the mixture, the temperature rise via heat transfer is reduced and thus the combustion rate is slowed down and peak pressures reduced. Also, the heat capacity of the charge before combustion rises with increased EGR.

Cairns et al. conducted a number of studies concerning the benefits of EGR against other knock control technologies like lean boost and enrichment [26, 27, 28]. They found EGR to be significantly more effective than air or fuel excess. Furthermore benefits in fuel consumption were measured up to 17 % with improvements in CO and CO_2 emissions. Alger et al. explored the effects of EGR on combustion efficiency and engine knocking at high engine load [4]. Findings were also a reduced fuel consumption at high engine load, attributed to eliminating the requirement for enrichment and reducing the knock tendency. Alger et al. additionally investigated the potential of multiple injections to improve the EGR tolerance and reduce the cycle-to-cycle variations. It could be observed that injecting 10 % to 15 % of the fuel mass late in the compression stroke resulted in an improvement in engine stability, burn rates and overall performance. The primary reason for this was an enhancement of the early burn rates, which is assumed to be caused due to locally rich areas around the spark plug.

The findings of Cairns et al. [26, 27, 28] and Alger et al. [5] confirm the previous work of Grandin and Ångström [64, 65], which also highlights the benefits of EGR allowing a stoichiometric operation across the full operation range of the engine.

EGR architecture

Using EGR, several different architectures are possible for internal combustion engines. These different architectures can be distinguished by the position from which the exhaust gases are sourced (pre- or post-turbine) and the location, where the EGR is mixed with the charge air (pre- or post-compressor). Vitek et al. published a simulative comparison of four different EGR-architectures: 'HP[11]', 'LP[12]', 'HP-LP' and 'LP downstream DPF' [194]. They state that LP architecture shows the potential to deliver the highest EGR-rates and therefore the highest potential to reduce fuel consumption under steady state conditions. For transient operation on the other hand, the HP configuration shows advantages in the response behaviour. Potteau et al. conducted a comparable simulative study and additional experimental work with a HP and a LP EGR system [151]. For HP EGR, higher EGR rates, especially

---

[11]high pressure
[12]low pressure

at low engine speeds, are only realisable by using a throttle in the exhaust path to increase the pressure differential between intake and exhaust manifold and thus to increase the EGR flow rate. Comparing the two EGR configurations to the non-EGR operation, Potteau et al. describe a 6.6 % BSFC benefit for HP EGR and a 13.6 % improvement for LP EGR for steady state operation at high engine speed (5500 rpm) and enhanced engine load (1.29 MPa BMEP). The difference in BSFC improvement can be explained by the higher knock-limited spark advance improvement using the LP configuration. This higher knocking resistance is attributable to the cooler EGR temperature. Furthermore, as the EGR was taken post catalyst, a reduction of HC and CO levels for the LP configuration shows potential to reduce engine knocking, as shown in the following section 2.4.5. Alger et al. investigated the potential of both cooled and uncooled HP EGR and LP EGR system [4] and state:

"The addition of EGR at low and part load was shown to decrease NO and CO emissions and to reduce fuel consumption by up to 4 %, primarily through the reduction in pumping losses. At high loads, the addition of EGR resulted in higher fuel consumption benefits of 10 % - 20 % as well as the expected NO and CO reductions. The fuel economy benefit at high loads resulted from a decrease in knock tendency and a subsequent improvement in combustion phasing as well as reductions in exhaust temperatures that eliminated the requirement for over-fuelling."

Turner et al. published a comparison of HP, LP and HP-LP EGR configurations [190]. The focus thereby was on the high load operation at high engine speeds (5000 rpm). Compared to other literature, they used relatively low EGR rates of about 6 % and found that the different architecture of the EGR system had a stronger impact on the charging system than on the combustion system. Also, LP EGR architecture appeared to offer the least benefits. However, they noted that the turbocharger was not matched well to the engine. With an optimised turbocharger, the benefits may have been significantly improved. Furthermore, the mismatch of the turbocharger could have different effects for each EGR configuration. In general, LP EGR was beneficial compared to HP EGR in terms of mass flow at low engine speeds, because the HP EGR rate was limited by the pressure difference between exhaust and intake. To increase the EGR-rate in HP EGR configuration, a further increase of the exhaust back-pressure would be needed, which would reduce the volumetric efficiency. LP EGR on the other hand showed a slower transient response.

Catalysed EGR

As EGR is a mixture of different molecules with different effects on the combustion process, a discussion of the most important involved components was done by Parsons et al. [144]. Even though EGR is often described as the introduction of inert gases to the inlet charge, Parsons et al. state that the chemistry of combustion is also affected by some of the components of the exhaust gases. Grandin and Ångström state that especially at high engine load "end gas reactions can be reduced either by lowering the temperature or by decreasing the concentration of radicals by chemical intervention" [65]. The most reactive parts of the exhaust gas components are nitrogen oxide, carbon monoxide and unburned hydrocarbons. By using a low pressure EGR (LP EGR) architecture, it is possible to use the exhaust gas downstream of the catalyst. This changes the chemical composition of the gases introduced to the inlet charge. Topinka et al. investigated the effect of H_2 and CO, both in isolation and combined, on the combustion process [187]. They state that "H_2 and CO inhibit knock by slowing autoignition chemistry and slightly increasing flame speed" [187]. For CO they estimated an octane number of 106, while in a previous study by Tang et al. [186], the octane number for H_2 was estimated to 130. This knowledge lead to investigations by Fennell et al. on reforming the recirculated exhaust gas through a catalyst with a fuel injection to produce hydrogen-rich gas at the inlet [49]. Improved combustion stability with reformed EGR was found by Fennell et al. during low-load operation, enabling higher EGR rates and further dethrottling.

Hoffmeyer et al. investigated the effect of catalysed EGR using a HP configuration and found an improvement in BSFC of about 2 % with catalysed EGR compared to non-catalysed EGR [85]. They attributed the benefit to the earlier combustion phasing of about 1.5 - 3 °CA , which could be reduced to the increased knocking tolerance. As the CO_2-content is increased by the catalytic reactions, the heat capacity is increased compared to the non-catalysed EGR. Furthermore the amount of NO and HC is reduced. As the effect of the various components of the EGR on the ignition delay was not clear, Hoffmeyer et al. conducted a sensitivity study for some of the EGR components. They found, that the reduced knocking tendency was driven by the reduction of NO and acetylene (C_2H_2). However, the possible error in the analysis because of the EGR-measurement should be mentioned. Hoffmeyer measured the EGR ratio conventionally by comparing intake and exhaust CO_2-ratios. By using a catalyst, the amount of CO_2 is increased. Hoffmeyer et al. calculated that the catalytic process introduced an error of less than 2 % in their study. To achieve BMEP values of up to 2.65 MPa, Lewis et al. compared both catalysed and non-catalysed EGR [119]. In contrast to the previous discussed publications, Lewis et al. operated the engine with a constant total charge air flow (air + EGR). Therefore, engine load and injected fuel mass decreased with higher EGR rates because of the reduced amount of oxygen and the engine being set to maintain stoichiometry. For equivalent EGR mass flows, they found an improvement of combustion phasing due to reduced knocking tendency of about 6 °CA for catalysed EGR compared to non-catalysed EGR.

EGR and particle emissions

Concerning soot emissions, the tendency for increasing soot emissions using EGR is known from Diesel engines [79]. For gasoline engines however, there are contradictory publications on the effect of EGR on PN emissions:
Alger et al. used a four cylinder engine with port fuel injection and a high pressure EGR system [5]. Using an EGR rate up to 25 %, the PN emissions were increased

for operation at medium load (0.67 MPa IMEP) and low engine speed (2500 rpm). At higher engine speed and load (3500 rpm and 0.98 MPa IMEP) however, the PN emissions could be reduced using the same amount of EGR (25 %). Thereby the average particle diameter was increased with higher EGR fraction.

Hedge et al. showed a reduction of PN emissions by using low pressure EGR with an inter-cooler in a four cylinder GDI engine [76]. Especially at knock-limited operation, the PN emissions could be reduced. The EGR reduces the peak temperature in the combustion chamber and thus prevents the engine from knocking. Therefore, the typically used enrichment to cool the engine is not necessary. Hedge et al. also compare the effect of internal and external EGR. Whereby internal EGR showed a higher potential to reduce PN emissions.

Sabathil et al. investigated the influence of different operation parameters on the PN emissions in a four cylinder engine with a centrally mounted, multihole injector [164]. No benefits concerning the use of EGR were measurable for this engine at low engine speed (1000 rpm) and low engine load (70 Nm). Using a smaller amount of EGR (up to 12 %), increased PN emissions were measured. By further increasing the EGR up to 20 %, the PN emissions could be reduced to the baseline level (without EGR).

Winkler et al. investigated a low pressure EGR system on a downsized four cylinder engine with 1.25 l displacement [205]. Their attempt was to reduce pumping losses at part load operation and to prevent the engine from knocking at high load operation using EGR. At part load operation, they showed a reduction of PN emissions for higher EGR rates. For knock-limited operation, the effect of increased EGR rate was a reduction of PN at constant MFB50%. As the EGR prevents the engine from knocking, advancing the spark timing was possible. Thereby the PN emissions were increased for earlier mass fraction burned 50% (MFB50%). The authors state that the reduction of PN at constant MFB50% could be attributed to the cooler combustion using EGR. However, no information on knocking intensities, exhaust back-pressure and charge air pressure as well as exhaust gas temperatures are given in the paper. The increased PN emissions at optimised MFB50% timings are a hint that the PN reduction at constant MFB50% was driven by the post-oxidation process. The authors state that the most important influence of EGR is the reduction of the knocking tendency and thus a possible reduction of enrichment for part protection.

Pei et al. investigated the influence of EGR at low engine load conditions (0.1 MPa and 0.3 MPa IMEP) [145]. Even though the authors do not comment on the EGR architecture, it is assumed that a high-pressure EGR setup was used. It should be taken into account that the engine was operated with a low injection pressure of 10 MPa and that the baseline PN concentration was on a high level[13]. For EGR rates up to 20 %, the concentration of accumulation mode particles can be reduced, whereas the concentration of nucleation mode particles is significantly increased. In summary, the total number count for both engine loads is increased using EGR. The authors state that the increased in-cylinder temperature with higher EGR rates improves the fuel evapouration and atomisation. Furthermore, the authors assume that the fuel im-

---

[13]$\approx 1.2 \cdot 10^8$ #/cm³ for 0.1 MPa IMEP and 0% EGR; $\approx 4 \cdot 10^7$ #/cm³ for 0.3 MPa IMEP and 0% EGR

pingement to the wall could be reduced, leading to a more complete combustion and inhibiting the generation of accumulation mode particles. They further state that the reduced maximum combustion temperature using EGR may reduce the pyrolysis and dehydrogenation of the fuel and inhibit the generation of accumulation-mode particles further. Bogarra-Macias et al. investigated the potential of reformate EGR (REGR) on particle emissions of a GDI engine [21]. A fuel reformer for hydrogen-rich gas was used in combination with a LP EGR system. Fennell et al. showed, that H_2 promotes OH production that can also slow down the rate of soot formation [49]. Additionally, Stone et al. showed that molecular hydrogen (H_2) enhances the rate of soot formation and oxidation by increasing the in-cylinder temperatures [182]. Bogarra-Macias et al. estimated the different effects of EGR and REGR on the combustion process, soot formation and soot oxidation. First of all, both EGR and REGR dilute the charge and thus reduce the overall oxidant concentration, which prompts soot formation and slows down soot oxidation. Secondly, EGR decreases the in-cylinder temperature and thus inhibits both soot formation and oxidation, while REGR increases the in-cylinder temperatures and thus promotes soot formation and oxidation. In their experiments, the soot reducing effects of REGR dominated and therefore reduced the soot emissions. Regarding chemical effects, it is believed that hydrogen addition inhibits soot nucleation by slowing or reversing the hydrogen abstraction acetylene addition mechanism for soot formation [143]. Saxena stated that H_2 also promotes the formation of the OH radical [167], which is effective for soot and soot precursor oxidation [36]. However, Bogarra-Macias et al. stated:

"[...] when CO is present in the combustion chamber it is oxidised by both oxygen and OH radicals, reducing the concentration of these species available for soot oxidation. As EGR and reformate EGR increase the engine efficiency, for a given load less fuel needs to be injected and thus less potential for soot formation is present. For reformate EGR, this effect is more noticeable because a part of the hydrocarbon based fuel is replaced by hydrogen, resulting in less locally fuel rich zones, where most of the soot is produced [168], and a reduction of the carbonaceous content inhibiting soot formation and further growth."

2.4.6 Fuel-composition

The fuel composition has a significant influence on the particle formation process, as already shown in section 2.2. Therefore, Aikawa et al. developed a model to correlate particle number emissions with the vapour pressure and the double bond equivalent (DBE) of the fuel components [2]. The DBE is defined as follows:

$$DBE = \frac{2 \cdot C + 2 - H + N}{2} \tag{2.4}$$

where C, H and N are the number of carbon, hydrogen and nitrogen atoms respectively in an organic compound.

The PM index defined by Aikawa links the PM emissions with the vapour pressure (V.P) and the double bond equivalent (DBE) of the components in the fuel weighted by mass fraction (Wt) as follows:

$$PMIndex = \sum_{i=1}^{n} I_{(443K)} = \sum_{i=1}^{n} \frac{DBE_i + 1}{V.P_{(443K)}} \cdot Wt_i \qquad (2.5)$$

However, it should be noted that an engine with port fuel injection was used and that no independent control of the fuel vapour pressure or DBE was possible, as commercial fuels were used along with a base fuel to which different components were added. Furthermore, Aikawa et al. evaluated the vapour pressure at a range of temperatures and found the best correlation between PM Index and particle emissions for the vapour pressure at a temperature of 443 K. The correlation of the PN emissions versus the PM Index of the fuels used by Aikawa et al. is shown in Figure 2.11 on the left hand side. The PM Index of the fuels used was between 1.01 to 3.86. The coefficient of determination is 0.95, showing a good model fit. However, it should be noted that fuel No. 9 (Base + Indene, C_9H_8) with a PM Index of 2.09 and PN emissions of $6.85 \cdot 10^{12}$ #/km was obviously defined as outlier and not used for the correlation.

Leach et al. calculated the PM Index for a selection of commercially available fuels [116], as shown in Figure 2.11 on the right hand side. The mean PM Index for these fuels is 2.12 and the standard deviation 0.81.

Leach et al. further investigated the model of Aikawa et al. using a DISI engine with designed fuels, whereas the DBE and the vapour pressure were varied independently [116]. In contrast to Aikawa et al., who calculated the PM Index by volume fractions, Leach et al. calculated a PN Index using mass fractions. However, they state that in the majority of cases, the relative difference in calculation of PM Index and PN Index is less than 15 %.

Summed up, Leach et al. state:

"The effect of low boiling point components on the fuel spray is significant, and the addition of pentane to fuels mixed from pure components is important in order to reflect real world evapouration behaviour. [...] Using these criteria, a recipe has been generated to create model fuels to mimic commercially available gasolines. A matrix of these model fuels has been tested [...] and their PN emissions have been shown to follow the PN index. [...] The PN index has been validated in a SGDI engine by using both model fuels and commercially available fuels. The PN Index has been shown to be an important parameter for fuel specification."

However, it should be noted that Leach et al. operated the engine at low engine speed (1500 rpm), low load (0.18 MPa IMEP) and with a rich mixture ($\lambda = 0.9$).

Aikawa et al. and Leach et al. thus showed the potential to reduce particle number emissions from GDI engines by changing the fuel composition. However, both their investigations show differences to advanced GDI engines in the market. Aikawa et

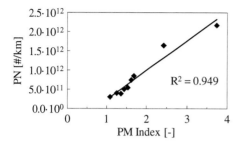 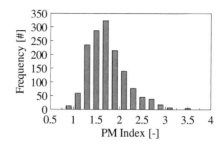

Figure 2.11: Left hand side: Correlation of PM Index and PN emissions by Aikawa [2];
Right hand side: Range of PM indexes of commercially available fuels worldwide by Leach et al. [116] with data from Eastwood [42]

al. used an engine with port fuel injection, whereas Leach et al. operated the engine
under fuel-rich conditions ($\lambda = 0.9$).

The non-stationary operation and the first phase of any test cycle are important concerning exhaust gas emissions. Till the catalyst light-off, a higher exhaust gas enthalpy
gets provided by the engine during the catalyst heating operation. To reduce the exhaust gas emissions and especially the particle number emissions, Dageförde et al.
investigated different fuels and fuel blends at catalyst heating operation [32]. The
results obtained by Dageförde et al. support the findings of Aikawa et al. and Leach
et al., as the particle number emissions could be reduced by operating the engine with
biogenic fuel blends E40 and B40[14] as well as with alkylate fuel[15]. These blends have
a reduced amount of aromatic compounds and thus a lower DBE. However, using
toluene as a neat aromatic compound and thus increasing DBE, did not increase the
particle number emissions. Dageförde et al. state that not only the aromatic fraction
but also the fuel properties influencing the spray break-up and mixture formation process are responsible for the particle formation process. It should also be mentioned
that two different injector types were used for the investigations: a multi-hole injector
and an outward opening nozzle. Dageförde et al. state that due to the different nozzle
types and spray targeting, the particle number emissions depend strongly on the injection strategy. However, the different injection strategies showed a higher potential
to reduce particle number emissions compared to the fuel composition.

Based on the results of Dageförde et al, Bertsch et al. used alkylate fuel at higher
engine load [16]. For these investigations, only the multi-hole injector and the RON 95
and alkylate fuel used by Dageförde et al. were used. The operating point was shifted
to a higher engine speed (2000 rpm) and higher engine load (1.4 MPa IMEP) and different charge motion strategies were used. A reduction of particle number emissions
was possible for almost all settings. The PN reduction potential was depending on

---

[14]E40 was RON 95 blended with 40 % ethanol, while RON 95 was blended with 40 % butanol for B40
[15]Special fuel designed with 0 % aromatic compounds but comparable physical properties to RON 95
such as octane number and evapouration curve.

the particle number concentration and higher for operating points with high baseline emissions. However, the implementation of a large scale charge motion showed a higher potential to reduce PN emissions. Additional investigations using a spectrograph showed that alkylate fuel did not tend to build up deposits in the combustion chamber and the optical accesses.

Maier et al. investigated the fuel-independent sources of particle emissions [121]. They operated a single cylinder research engine with regular RON 95 in the first step and switched to operation with methan (CH_4) in the second step, before finally operating with hydrogen (H_2). Concerning the particles with a diameter larger than 10 nm, the number concentrations could be reduced by the substitution of carbonaceous fuel with hydrogen. However, concerning the sub-10 nm particles, the number concentrations remained almost constant. As the authors used a PMP compliant setup for the particle measurement with a VPR, it is assumed that these are non-volatile particles. The authors also excluded these sub-10 nm particles to be caused of high ambient concentrations, as the particle load of the inlet air was measured and found to be orders of magnitude below the concentrations in the exhaust. Therefore, the authors assume these particles to be caused by abrasion from the cylinder liner or metal additives from the lubricating oil.

3 Experimental setup

3.1 Test bench setup

The test bench used is equipped with several measurement devices, as shown in Figure 3.1 [16, 32, 75]: A Pierburg PLU 401 measures the fuel consumption. The exhaust A/F ratio is measured by using an ETAS LA,4 Lambda Meter. For exhaust gas emissions a Pierburg AMA 2000 (THC, NO_x, CO, CO_2, O_2) is used.

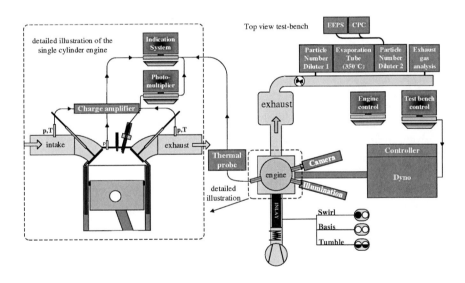

Figure 3.1: Test bench setup

Besides the gaseous exhaust emissions, the particle number emissions are measured by using the following setup: A chamfered sample probe is mounted and connected to an AVL 489 Particle Counter Advanced [56]. The short (0.5 m) sample line between probe and primary dilution (Chopper diluter) is heated to 190 °C. The following evaporation tube and second dilution stage are connected to the chopper diluter via a heated line.

These three devices combined form the *Volatile Particle Remover* (VPR) that operates according to the guidelines of the *Particle Measurement Programme* (PMP), as discussed in section 2.3. The overall dilution can be adjusted between 100:1 and 2000:1 depending on the particle number concentration in the exhaust gas. Downstream the VPR the conditioned sample gas is split. A part of the gas stream is fed to the TSI 3790 *Condensation Particle Counter* (CPC) of the AVL device. The remaining part of the diluted gas is fed to a TSI 3090 *Engine Exhaust Particle Sizer* (EEPS) [189]. The CPC fulfills the requirements of the UNECE-R83 with a counting efficiency of 50 % for particles with 23 nm ($\pm$ 1 nm) electrical mobility diameter. The maximum measurable particle diameter is 2.5 μm. The EEPS is not conforming to UNECE-R83, but it is attached to complement the measured particulate number concentration of the CPC with information about particle size distribution between 5.6 nm and 560 nm electrical mobility diameter.

The in-cylinder pressure is measured using a Kistler 6054AR piezo-electrical pressure transducer. In combination with the low-pressure piezo-resistive transducers (Kistler 4045 and 4075) mounted in intake and exhaust ports, the pressure signals are the basis for a GT-Power TPA model of the engine, as shown later in section 3.3. For each measurement, 500 consecutive cycles with a temporal resolution of 3,600 pulses per revolution are acquired.

For the investigations with exhaust gas recirculation (EGR), a high-pressure EGR-system including an inter-cooler was applied to the engine. The setup of the system is shown in Figure 3.2. The exhaust gas was extracted shortly after the exhaust runners and before the exhaust back-pressure throttle, which simulates the turbine of a turbocharger at the single cylinder test bench. The secondary circuit of the inter-cooler used could be operated with engine cooling at 90 °C or with cool water at 20 °C respectively. Operation with engine cooling is thereby representing actual setups in automotive applications, whereas the operation with cool water is used to show the potential of cooled EGR.

The EGR ratio was calculated at the test bench by measuring the concentration of carbon dioxide in the exhaust gas and charge air, as shown in equation (3.1).

$$EGR_{rate} = \frac{CO_2^{Intake} - CO_2^{Background}}{CO_2^{Exhaustgas} - CO_2^{Background}} \tag{3.1}$$

3.2 Engine data

The engine used for the investigations is a single cylinder research engine. Technical data of the engine is shown in Table 3.1. The engine is equipped with cam phasers both on intake and exhaust side and a centrally mounted injector. The spark plug and the injector are mounted in the longitudinal direction of the crankshaft. For the investigations a constant engine speed of 2000 rpm is chosen and the relative air-fuel

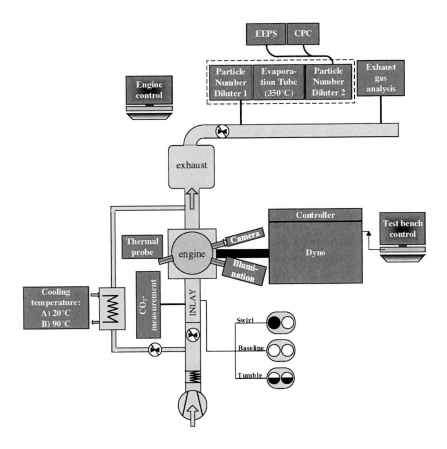

Figure 3.2: Engine test-bench with exhaust gas recirculation

ratio is set to stoichiometry ($\lambda = 1$). The exhaust gas back-pressure is set to a value of a reference full scale engine and changed equivalent with the charge air pressure for the outlined variations.

The cooling water temperature as well as the oil temperature is kept constant at $90°C$.

The base engine was designed for natural aspiration with low in-cylinder charge motion. At high engine loads, a high fuel mass needs to be injected. In combination with the lower in-cylinder charge motion caused by the low engine speeds, the mixture preparation is challenging. To improve the in-cylinder mixture formation, different strategies can be applied, such as optimised valve timings as exemplary shown by Tan et al. and Xu et al. [184, 207] as well as the implementation of a large scale charge motion, exemplary shown by Tang et al. [185]. The above-mentioned publications

Table 3.1: Engine data.

Displacement	[cm^3]	498
Stroke	[mm]	90
Bore	[mm]	84
Compression ratio	[-]	10.5:1
Intake valve timing variability	[°CA]	60
Exhaust valve timing variability	[°CA]	60
Max. intake valve lift	[mm]	0.5 - 9.7
Max. exhaust valve lift	[mm]	9.7
Injection system	[-]	centrally mounted DI

show that increased in-cylinder charge motion results in a more effective combustion process and hence shows the potential of reducing the particle number emissions. Different strategies to enhance the mixture formation can be applied. Firstly, a reduction of the intake valve lift can accelerate the flow velocities in the valve seat gap and thus enhance the mixture formation. A second strategy is the implementation of a large scale charge motion. Therefore, special inlays were designed for the intake port to generate a large scale charge motion [154]. To generate a sufficient tumble motion the lower parts of the intake ports is closed. Three different tumble-generating inlays were applied to the experimental setup: with 50 %, 60 % and 70 % of the intake ports area closed. On the right hand side of Figure 3.3 the swirl inlay is shown. The left inlet port is closed. The intake valve of the closed port is operating, but no charge air is induced via this port. The geometry of the inlays in the cylinder head is shown in Figure 3.3.

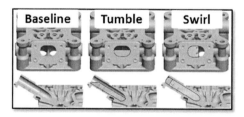

Figure 3.3: Generation of large scale charge motion: Baseline (left), Tumble (centre) and Swirl (right) [154]

The engine is equipped with cam phasers both on intake and exhaust side, making it possible to vary the intake valve spread to compare the charge motion strategies. By changing the intake valve spread, using a phase shifter, the intake open timing (ivo) as well as the intake closing timing are changed. At full lift, a later intake open timing leads to an intake closing event after the bottom dead centre. Thus the effective compression ratio is decreased. At lower intake valve lifts, the intake closing time is before the bottom dead centre. In this way, a reduction of the valve spread increases the effective compression ratio. This behaviour is shown in Figure 3.4. It is to mention that these values are theoretical and do not take into account any dynamic effects. For the engine used and maximum intake valve lift, the geometric compression ratio only equals the effective compression ratio for an intake open timing of about 325 °CA $_{aTDCf}$. Using a reduced maximum intake valve lift, the intake open timing needs to be close to or even after the top dead centre to reach the geometric compression ratio.

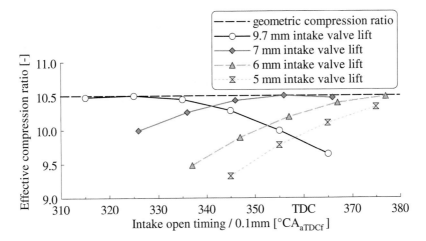

Figure 3.4: Effective compression ratio depending on the intake open timing

3.2.1 Ignition systems

For most of the investigations shown in the results section, a state-of-the-art transistor coil ignition was used with coil energies exceeding 55 mJ (dwell time set to 3 ms). The spark plug had a single-electrode and was additionally equipped with six optical fibres. It is shown in installed position on the right hand side of Figure 3.5. The observation fields of the optical fibres were directed to the piston (axial), comparable to the setup of Beck et al. [9]. When the flame front reaches an observation field, flame or soot radiation is transferred via an optical fibre to a photomultiplier which converts the detected radiation into a corresponding electrical signal [142]. The post-calculation process of the signals from the spark plug with fibre optical access (FO-SP) is shown in section 3.4.4.

To enhance the inflammation process, a high frequency ignition system (HFI), generating a corona discharge, was used. The system used for these investigations was a series of A-sample from BorgWarner Ludwigsburg GmbH, with the brand name EcoFlash, as introduced by Rixecker et al. [161]. It consists of a control unit and igniters, substituting the spark plug and pencil coil. Since the last step of voltage up-conversion is accomplished by a serial resonance circuit integrated in the igniter, the connector voltage stays limited to the sub-kV range [75]. The front-end electrodes of the igniter were formed to a regular star with five equally shared tips, as shown in Figure 3.5 on the left hand side. The tips of the igniter have to be sharp to generate a highly inhomogeneous electric field. Further information on the ignition system used is given by Bohne et al. [23] and Hampe et al.[74, 75].

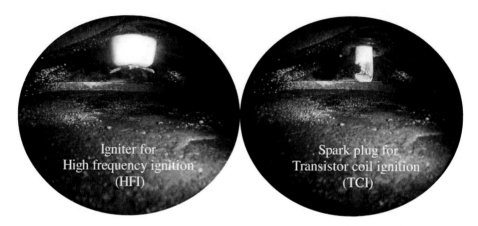

Figure 3.5: Ignition systems used [75]

3.2.2 Fuel properties and injector specifications

For most of the investigations a standard RON 95 gasoline fuel with 5 % ethanol content was used. The fuel parameters are shown in Table 3.2. The spray characteristics such as the spray cone angle, penetration length and droplet size spectra change depending on the injection pressure and the fuel characteristics. As shown in section 2.4.6, the PN emissions are strongly influenced by the fuel characteristics. Thereby the physical fuel properties, such as the surface tension and the viscosity, influence the injection and evaporation process. The chemical properties, such as the molecular composition and the vapour pressure influence the combustion process and the emission formation. Alkylate fuel is based on crude oil just like RON 95. However, the chemical composition is different, as the aromatic compounds were minimised close to 0 %. Alkylate fuel was designed for off-road applications, where the user of the hand-held machinery is directly exposed to the exhaust gases. Small two-stroke SI engines, which are mostly used in these applications, emit high amounts of unburned hydrocarbons. Using RON 95 thereby is more dangerous to human health because of the high amount of polycyclic aromatic hydrocarbons [13].

Alkylate fuel was designed to meet most of the specifications of RON 95, such as the evaporation behaviour, the caloric value and the octane number. RON 95 can be replaced by Alkylate fuel in most applications without changing the engine settings. The evaporation curves of RON 95 and Alkylate fuel are shown in Figure 3.6.

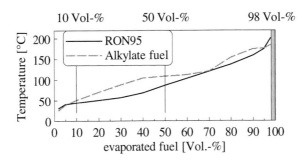

Figure 3.6: Fuel evaporation curves [32]

Because of the lower C/H-ratio, using Alkylate fuel the emission of CO_2 is lower compared to RON 95.

Table 3.2: Fuel properties.

		RON 95	ALKYLATE FUEL
Density*	[g/cm³]	0.75	0.7
Caloric value	[MJ/kg]	42.35	44.55
RON	[-]	95	95
MON	[-]	85	>90
Sensitivity	[-]	10	<5
Vapour pressure**	[kPa]	50-65	45-100
Kin. viscosity**	[10^{-6} mm²/s]	0.53	0.53
Surface tension**	[mN/m]	20	20
Evaporation heat*	[J/g]	350	210
Carbon mass fraction	[%]	84.25	84.19
Hydrogen mass fraction	[%]	13.51	15.81
Oxygen mass fraction	[%]	2.24	0
Aromatic volume fraction	[%]	28.3	<0
C/H-ratio	[-]	6.24	5.33
Stoichiometric AF	[-]	14.27	15.1

* at 20°C
** based on published data

Two injectors are used for the investigations: One with a high static flow of 820 g/min and one with a reduced static flow of 620 g/min. The reduced flow was designed to inject the same fuel mass at 35 MPa compared to the injector with the higher flow at 20 MPa and the same energising time. The two injectors match this design criteria well, as it is shown in the measurement of the injector characteristics measurements plotted in Figure 3.7. In the upper graph the injector characteristics of the low flow injector is shown for three different injection pressures (20 MPa, 35 MPa and 50 MPa). In the lower graph the same is shown for the high flow injector. Accentuated by the horizontal line at 25 mg/stroke injected fuel mass, it is obvious that the low flow injector using 35 MPa has a similar energising time as the high flow injector using 20 MPa.

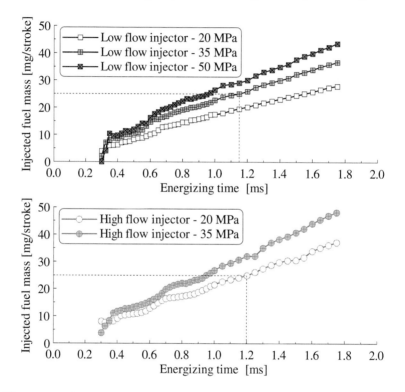

Figure 3.7: Characteristics of the injectors depending on the injection pressure

The number of holes and the spray pattern are equally laid out. Both injectors are solenoid activated and commercially available (Delphi Multec 14 and Bosch HDEV 5.2). Unlike the measurements done by Schumann et al. [174] using a Diesel setup to generate injection pressure up to 100 MPa, the injectors used were designed for operation with gasoline fuel. The spray targeting is adapted to the single cylinder research engine. For these investigations the engine is operated with a single injection in the intake stroke resulting in a homogeneous operation. The technical data of the injectors are shown in Table 3.3.

The average droplet size of the spray depends on the injection pressure, the back-pressure, the hole diameter as well as the fuel characteristics. Hiroyasu et al. found a formula to empirically approximate the *Sauter Mean Diameter* (SMD) [82, 81]. The formula is shown in equation (3.2). The investigations by Hiroyasu focussed on Diesel fuel. Aleiferis et al. tried to adapt the equation to gasoline fuel [3]. They showed, that the trends predicted by this equation are correct for gasoline fuel, while the absolute value of the SMD is underestimated by 50 %.

$$SMD = 0.38 \cdot d \cdot Re^{0.25} \cdot We^{-0.32} \cdot \left(\frac{\eta_a}{\eta_f}\right)^{0.37} \cdot \left(\frac{\rho_a}{\rho_f}\right)^{-0.47} \tag{3.2}$$

Schneider used the findings of Hiroyasu et al. as well as other literature to find proportionalities to describe the changes of the SMD [171]. This proportionality is shown in the following equation (3.3).

$$SMD \propto d^{0.93} \cdot \Delta p^{-0.2} \cdot \sigma^{0.32} \cdot \rho_f^{0.15} \cdot \rho_a^{-0.02} \cdot \eta_f^{0.12} \cdot \eta_a^{-0.37} \tag{3.3}$$

Using this proportionality under the assumption that the surface tension, the density for both gas and fuel and the viscosity for both gas and fuel stay constant during the injection at constant engine load and speed, it is possible to derive the relative SMD with different injection pressures and different injector flows. The results of this calculation are shown in Figure 3.8. The SMD of the high flow injector using 20 MPa injection pressure is set to 100 % and the reduction of the SMD by reducing the bore hole diameter (low flow injector) and increasing the injection pressure is shown.

By reducing the bore hole diameter, the SMD is reduced by about 15 % using 20 MPa injection pressure. With the reduced static flow, the injection duration needs to be extended to inject the same fuel mass using the same injection pressure. Increasing the injection pressure shortens the required injection duration and thus increases the time for the mixture preparation. By increasing the injection pressure, the SMD is reduced. Because of the specifications of the two injectors, the low flow injector using 35 MPa

Table 3.3: Injector properties.

		HIGH FLOW	LOW FLOW
Injector type		Mutli-hole injector	
Activation		Solenoid activated	
$\dot{Q}_{stat}$ [**]	[g/min]	820	620
Max. pressure	[MPa]	35	50
Spray jets		6	6
Spray angle	[deg]	75	75

[**] at 100 bar using n-heptane

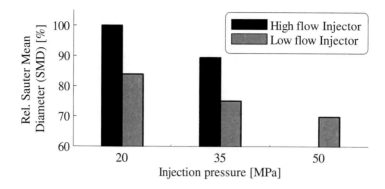

Figure 3.8: Relative SMD due to reduced bore hole diameter and increased injection pressure

injection pressure and the high flow injector using 20 MPa show the same static flow, as shown in Figure 3.7. Thereby the low flow injector injects smaller droplets with a 25 % reduced SMD compared to the high flow injector.

However, by reducing the bore hole diameter and thus the static flow, the maximum flow rate of the injector is limited. The injection duration needs to be prolonged at constant injection pressure to inject the same fuel mass. This reduces the mixture formation time until ignition timing. To compensate this effect, a higher injection pressure can be chosen. This results in disadvantages, for instance a higher fuel consumption caused by the higher required power of the fuel pump. For this work at a single cylinder test bench, the high pressure pump was driven externally and the higher required power for the pressure generation was disregarded.

3.3 Numerical analysis

The numerical analysis is essential to evaluate a combustion process. The pressure trace analysis provides information about the combustion process, while the gas-exchange calculation provides information about the mixture composition, especially the residual gas fraction, in the combustion chamber. For the numerical analysis of the experimental work, the software GT-Power[1] by Gamma Technologies Inc. was used. The tool Three Pressure Analysis (TPA) combines a pressure trace analysis with a gas-exchange calculation and a work cycle calculation. Besides the in-cylinder pressure signal, the software needs the pressure traces of intake and exhaust port. To get information about the wall temperatures, the optical accesses on the drive belt side was equipped with a thermal probe. Based on experiments by Hügel et al. [86],

---

[1]Version 7.3.0

this thermal probe consists of a steel plug with two fast response temperature sensors. Thereby one of the sensors is restored to measure the wall temperature while the second sensor sits flush with the combustion chamber wall, thus measuring the dynamic temperature profile of the engine.

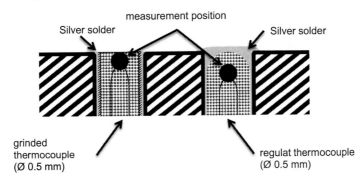

Figure 3.9: Thermocouples used in the optical access as thermal probe [33]

Within this work, only the function with averaged cycles was used (TPA steady). The engine was simulated between the intake and exhaust pressure transducers, which are mounted close to the intake and outlet respectively. The model for the engine was generated and applied with the help of a master student, Philipp Werner [199]. Further information on GT-Power and the TPA are given in [53, 54].

Using the conservation equations for mass, momentum and energy in flow direction, the gas-exchange calculation is solved for the parts through which the gas flows, resulting in averaged values for the flow area. The zero-dimensional[2] pressure trace analysis uses the first law of thermodynamics and the thermal state equation for ideal gases [79]. Thereby the wall heat losses need to be considered by models and the air-fuel equivalence ratio is needed as initial condition [44]. The instantaneous volume of the combustion chamber can be calculated via the engine geometry and crank angle information. A two-zone model to determine the mass fraction burned profile from the measured cylinder pressure data is used. Of special interest for this work are the start of combustion (MFB05%), the centre of combustion (MFB50%) and the end of combustion (MFB90%). Additionally calculated with these mass fraction burned points is the flame development angle (FDA), which is the time needed for the inflammation of the mixture and which is determined by the time difference of ignition timing and MFB05%. It is also possible to calculate the rapid burn angle (RBA), which describes the time of the cycle with a high mass fraction burn ratio. The RBA is determined by the time difference between the start of combustion (MFB05%) and the end of combustion (MFB90%).

The temperature trace inside the combustion chamber is also calculated. The two-zone model calculates a temperature for each burned and unburned zone. The burned

---

[2]Thermodynamic state values change in terms of time but not spatially

zone consists of the residual gas of the previous cycle before the start of combustion and increases during the combustion process, while the unburned zone shrinks. It is assumed, that the gases in each zone are homogeneously mixed and that the two zones are divided by an indefinitely thin reaction zone [79].

3.4 Optical diagnostics

As described in section 2.2, the particle formation processes are mostly a local phenomena which cannot be fully understood by global analysis like engine-out emission measurement or in-cylinder pressure trace analysis. However, optical diagnostics are well suited to provide in-situ, spatially resolved information on the soot formation process in combustion engines. Coupling the optical information with the in-cylinder pressure traces and the exhaust gas analysis is a promising solution to understand the causes of particle emissions from GDI engines.

3.4.1 Mie-scattering

Using a high-speed camera, different information about the mixture formation process and the combustion process can be visualised. Additionally, using a light source for illumination, the liquid phase of the injected fuel spray is visible according to the Mie-Scattering theory. The theory of Gustav Mie about the elastic scattering of light on spherical particles was first published in 1908 [134], and is described in detail in the work of van de Hulst [87]. Henle describes, that getting information of the spray-formation by a quantification of the scattered light is not possible in fuel sprays [78]. He describes that a quantification according to the Mie-Theory is valid only for light scattering on spherical particles, which are spread in a homogeneous medium and under the assumption that the particle distance in relation to the wavelength of the scattered light is large. Several limitations need to be applied, using the Mie-Theory on a fuel spray. The intensity distribution of the scattered light depends on the geometric shape of the droplets, because the droplets are assumed not to be of spherical shape. Furthermore, a large number of droplets with differing size, number and shape, which may have a very small distance exist in a fuel spray. Therefore, multiple scattering, reflections and partially extinction of the incident and scattered light influence the intensity distribution. Thus it is not possible to quantitatively calculate the injected fuel mass by analysing the scattered light intensity. Using the Mie-theory in combustion engines, additional problems like deposit formation on the optical accesses and reflections of the combustion chamber walls inhibit a quantitative calculation of the scattered light.

3.4.2 Radiation of flames

Energy is emitted by the working fluid in terms of electromagnetic radiation of the visible and near infrared spectrum in combustion engines. This radiation is either absorbed or reflected when hitting the combustion chamber walls. The absorbed radiation is converted to thermal energy and increases the wall temperature. The extent of radiation is determined by the emissions behaviour of the working fluid [149]. The radiation is an addition of the gas radiation and the soot luminosity.

Gas radiation in gasoline engines is mainly caused by the radiation of the gases H_2O, CO_2 and CO. They absorb and emit radiation almost exclusively within certain wavelength bands characteristic of each species. Therefore, they are also known as selective radiators.

The second source of radiation in gasoline engines is the soot luminosity, which emits light over a wide range of wavelength, similar to a 'grey body radiator'. Soot luminosity overpowers the intensity of the gas radiation. In combustion processes with locally fuel-rich conditions, soot is emitted caused by the diffusive combustion under oxygen deficiency [20], as shown in section 2.2. The broadband, thermal, solid state radiation is also known as soot luminosity and emits, depending on the temperature, radiation in the visible wavelength [55]. The intensity of the soot luminosity increases with higher soot particle number, soot particle size and their temperature. In gasoline engines, soot luminosity was not a common phenomenon for PFI engines, because of the good homogenisation by the early mixing of fuel and air. For gasoline direct injection, the time for mixture preparation is shorter. Therefore, it is possible that locally fuel rich zones exist in the gas phase, even after the start of combustion. Also, an impingement of liquid fuel on the piston surface, an interaction of the spray with the open inlet valve or a wetting of the liner and the combustion chamber roof is possible. Another possible source of soot emissions in GDI engines with multi-hole injectors is tip-sooting, as already described in section 2.4. However, most of these soot emissions can be visualised using a camera system, as shown by Steimle et al. [180], and shown in Figure 3.10.

3.4.3 Adaption of the high-speed measurement equipment

Optical sleeves with sapphire windows can be mounted in each of the four optical accesses of the research engine. For the optical investigations, one of the accesses is used for illumination and one is used for a CMOS camera (LaVision HSS6). The high-speed camera operates with the corresponding commercial software DaVis 8.3. It has a maximum resolution of 1024 x 1024 pixels (20 μm per pixel). Due to the chosen high repetition rate of 13.5 kHz, the camera chip is reduced to an area of interest of about 512 x 512 pixels to lower the memory required for the measurements. The camera-timing and the crank angle-timing are synchronised using a hypersampling setup and recorded using the indication system. For every operating point 300 cycles are recorded from -310 °CA$_{aTDCf}$ to 90 °CA$_{aTDCf}$. The repetition rate of 13.5 kHz

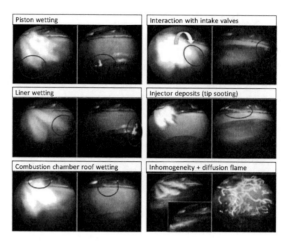

Figure 3.10: Some of the most important sources of soot emissions in GDI engines according to Steimle et al. [180]

results in a temporal resolution of $\approx 1\,°\text{CA}$ at 2000 rpm. For the illumination, a cold-light source with a 270 W metal halide lamp is used (Storz Techno-light 270). The setup of the optical accesses and the field of view in the combustion chamber are shown in Figure 3.11. Besides the optical accesses, the intake and exhaust valves are visible as well as the spark plug and the centrally mounted injector.

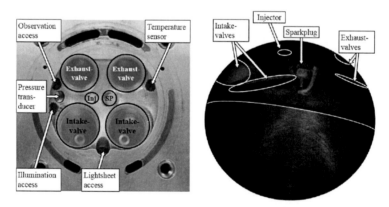

Figure 3.11: Setup sensors in the combustion chamber (left) and field of view of the optical access (right)

Due to the deposit formation during fired engine operation, it is not possible to drive the engine for a long time until all temperatures become steady. Deposits reduce the transmission of light through the optical access and thus reduce the information gathered by the high-speed camera measurements. However, to measure under re-peatable conditions, the engine was motored at 2000 rpm until the engine cooling and the oil reached the target temperature of 90°C. After these criteria were reached, the boost pressure was increased up to the required value. Then ignition and injection were enabled and ran for about one to two minutes, before starting the high-speed measurement. This procedure showed the best compromise between reproducible con-ditions and the formation of deposits on the optical accesses.

As Köpple et al. showed for a load step from 0.2 MPa IMEP to 0.8 MPa IMEP at 2000 rpm, the temperature of the piston surface needs about 100 seconds of time to reach the higher temperature level [110]. The piston surface temperature is the most important value concerning fuel impingement for a GDI engine with a centrally mounted injector. Köpple et al. state that caused by the very low temperature of the piston at the start of the load step, the fuel deposited on the piston surface can not evaporate completely. This results in fuel-rich zones close to the piston and a resulting diffusive combustion causing high particle number emissions. As the higher tempera-ture of the increased engine load reached the steady state level, a substantially smaller amount of fuel impinged on the piston surface is stored as wall film. And because of the higher temperature, the deposited fuel evaporates before the onset of combustion, thus resulting in a close to premixed combustion with low particle number emissions.

3.4.4 Spark plug with fibre optical access - post-processing

Six fibre optical probes were radially applied to a spark plug around the circumference directed to the piston (axial). When light is emitted in the observation area of the fibre optics, either by gas or soot radiation, the radiation is transferred via the fibre optics to a photomultiplier, which converts the detected radiation into a correspond-ing electrical signal. These electrical signals are transferred to the indication system and measured with the resolution of 0.1 °CA in phase with the other indicated signals. There has been research done using a Spark plug with fibre optical access to determine the soot emissions of GDI engines [83, 165]. Hirsch et al. used a spark plug with one fibre optical probe and an integrated pressure transducer to calibrate a GDI engine in transient operation [83]. They found out that for homogeneous GDI engine operation, the luminosity of the flame contributing to the power output is essentially synchronous to the rate of heat release and that any significant difference between flame luminosity and rate of heat release is an indication for imperfect mixture formation. Finally they state that "diffusion flames in gasoline engines are either the signature for an articulate stratified combustion, or they are the result of insufficient mixture formation. Finding diffusion flames at nominally premixed combustion conditions, thus is evidence for the necessity to improve mixture formation" [83].

Sabathil et al. used a Spark plug with fibre optical access (14 fibre optical probes) and a camera system (CCD) to find explanations for the PN concentrations measured [165]. They firstly used the fibre optics and the camera system as stand-alone and later on in combination to show the efficiency of each system in finding sources of PN emissions. The authors state that most information on soot formation can be gathered by a combined usage of the spark plug with fibre optical access and the camera system, although most effort is needed for the combined usage. They confirm the findings of Hirsch et al. in measuring the soot luminosity with the spark plug with fibre optical access.

Schwenger et al. described different characteristic values for the post-processing of optical signals to correlate the emitted radiation to measured exhaust gas concentrations [175]. Schwenger et al. used two-colour pyrometry and correlated the emitted values of a Diesel engine to the measured FSN values. The characteristic values, for instance the position of the peak intensity, the value of the peak intensity or the gradient of the intensity, were tested to correlate the radiation signals of the spark plug with fibre optical access to the measured particle number concentrations. The measured signals of the fibre optics for one representative cycle for SOI = -300 °CA $_{aTDCf}$ (left hand side) and for SOI = -340 °CA $_{aTDCf}$ are shown in Figure 3.12. In the upper graphs, the signals of the fibre optics are plotted and in the lower graphs the pressure traces are plotted versus the crank angle. Even though the pressure traces appear identical, the fibre optical signals are significantly different. Operating with a SOI of -300 °CA $_{aTDCf}$, the PN emissions are on a low level and the fibre optical signal shows the radiation of the regular combustion. Advancing the injection timing leads to an impingement of liquid fuel on the piston surface, which can not be evaporated until the start of combustion, leading to a diffusive combustion on the piston surface ('Poolfire'). The diffusive combustion becomes visible as the fibre optical signal shows increased values after the regular combustion process.

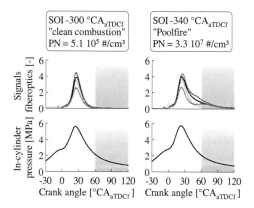

Figure 3.12: Post-processing of fibre optical signals

The post-processing of the fibre optical signals focussed on the late expansion stroke, as the diffusive combustion was visible and distinguishable from the regular combustion. The maximum signal intensity of the fibre optics depends on the maximum light emitted in the observation area of each probe, the gain of the photomultiplier and strongly on the deposits formed at the optical probe. Therefore, the intensity of each fibre optical signal was normalised for the post-processing. Thereafter an integration of the intensity was performed from $60\,°CA_{aTDCf}$ until $120\,°CA_{aTDCf}$, as shown in the highlighted box in Figure 3.12. The mathematical description is shown in equation 3.4.

$$Soot\ integral = \int_{60}^{120} \frac{Intensity(\alpha)}{max(Intensity)}\ \mathrm{d}\alpha \qquad (3.4)$$

3.5 Particle Image Velocimetry

The in-cylinder flow field is influenced by different issues as shown in section 2.4.2. To determine the in-cylinder flow field and the interactions of the fuel injected and the in-cylinder charge motion, optical measurement techniques like Particle Image Velocimetry (PIV) can be applied. The PIV setup used in theses investigations was based on the investigations by Disch et al. [39]. As the focus was on real engine conditions, the PIV system was setup to operate under fired engine operation.

Optical diagnostics are commonly used for the qualitative and quantitative investigation of flows in technical applications. Kohse-Höinghaus and Jeffries published an overview on flow field measurement techniques [107] and Raffel et al. published a practical guide regarding the particle image velocimetry technique [153]. The theoretical limits of PIV were investigated by Keane and Adrian [98], Adrian [1] as well as Westerweel [200, 201].

Due to the high relevance of the in-cylinder flow field on the mixture formation and combustion process of internal combustion engines, a lot of studies have been published focussing on PIV measurements in either fired or motored engine conditions. Towers and Towers showed the possibility to measure the cycle-to-cycle variations of a combustion engine using high-speed PIV with a camera frame rate (cfr) of 13.5 kHz [188]. Stansfield et al. showed the possibility of measuring the in-cylinder flow field in combustion engines at motored operation with engine speeds up to 3500 rpm [177]. They showed that the higher fluctuations of in-cylinder flow dynamics and pressure wave fluctuations in inlet and exhaust port at higher engine speeds lead to inhomogeneous particle density distributions and thus to challenges for the PIV measurements. Müller et al. used PIV to determine the average and turbulent kinetic energy of the flow field in a motored combustion engine [139]. The high-speed setup used with a cfr of 6 kHz at an engine speed of 1000 rpm resulted in a temporal resolution of about one degree crank angle. Disch et al. used high-speed PIV to determine the instantaneous in-cylinder flow field of a DISI engine in stratified operation [39]. They used a cfr of up to 36 kHz which allowed measurements with a temporal resolution of $0.5\,°CA$ up to an engine speed of 3000 rpm. Disch et al. investigated the interaction of multiple injections and the in-cylinder tumble flow at fired engine operation.

3.5.1 Setup PIV

The PIV setup at the test bench is schematically shown in Figure 3.13. The optical accesses shown in Figure 3.11 were used for the PIV setup. Due to the small dimensions of the optical accesses ($\varnothing$ 8 mm), it is possible to run the engine continuously, ensuring the transferability of the results to conventional engines. On the other hand, the small diameter limits the maximum dimensions of the light sheet and thus directly the maximum detectable flow field. As the optical access for the light sheet, shown in the left graph of Figure 3.11 on the bottom, is not horizontal to the piston surface but with an angle of about 15°, the light sheet is oriented to the piston and exposes neither the spark plug area nor the injector. However, PIV was used in this project to show the influence of the inlays to generate a large scale charge motion inside the cylinder, which is assumed to be possible even with the light sheet tilted towards the piston. The application of endoscopes for the illumination and the recording in PIV systems was investigated and validated by Dierksheide et al. [38].

For the investigations, an Ar^+ laser[3] (cw) was used. The laser has a maximum light power of 5 W and emits light in the blue-green range (main wavelength are 488 nm and 514.5 nm). The laser beam was formed into a laser light sheet with a thickness of about 1.5 mm and optics that were adapted in the optical access. The settings of the high-speed camera were already shown in section 3.4.3. The cfr was increased up to 24 kHz for the PIV investigations to ensure a temporal resolution of 0.5 °CA at 2000 rpm engine speed.

The particles used, which are more closely described in the next section, were seeded in a cyclone and fed to the charge air after the charge air cooler. The high-speed camera was operated with a controller and synchronised to the indication system.

3.5.2 Particles used for PIV

Particle image velocimetry requires the use of seeding particles. The velocity of the particles thereby has to be representative for the flow field. Therefore, the shape, density and size of the particles should be chosen to have little momentum and thus being able to follow the instantaneous flow field. Studies on particles for PIV in fluid mechanics were done by Melling et al. [133] and Sankar et al. [166]. A typical trade-off for particles in PIV applications is the light scattering intensity, which increases with larger diameters versus the flow tracking characteristics, which are improved for smaller diameters. Different investigations showed a good compromise for the use of hollow thin-shelled polymer microspheres, which are filled with butanol, for engine applications [17, 39, 188]. These particles have a diameter of about 20-40 µm and a density of 20 kg/m^3. Disch et al. [39] described the challenge of particle seeding in combustion engines:

"The challenge for engine applications is to provide the correct number density of

---

[3]Type: Coherent Innova 70C

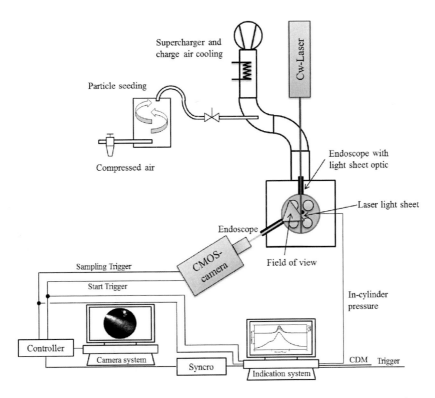

Figure 3.13: Setup for PIV measurements at the single cylinder test bench

seeding particles during the recording time due to ever changing spatial and temporal conditions inside the combustion chamber. Because of the continuously decreasing cylinder volume and the increasing particle density, it is challenging to achieve the optimal particle density during the late compression stroke."

The suspension of the particles is schematically shown in Figure 3.13. To prepare the solid microspheres in a non-agglomerated status, the particles were prepared in a cyclone generator. Due to the swirl flow inside the cyclone, caused by the tangential feed of compressed air, the particles were homogeneously distributed. The quantity of particles supplied to the engine could be adjusted by the pressure level of the compressed air.

To ensure reproducibility and comparability of the measurements, a procedure like for the high-speed measurements was performed. Thereby the engine was motored up to the nominal engine speed of 2000 rpm, highlighted as (1) in Figure 3.14. As the desired temperatures of oil and cooling were reached, the fuel injection and ignition were enabled, highlighted as (2) in Figure 3.14. When the stability of the exhaust gas emissions and the combustion behaviour were reached, the particles were fed to the intake air, highlighted as (3) in Figure 3.14. Shortly thereafter, the measurement with the high-speed camera was started. Thereby a steep increase in hydrocarbon emissions, and PN emissions was observed, along with a reduction of nitrogen oxide emissions and the relative air-fuel ratio. This behaviour is assumed to be caused by the combustion of the microspheres. As the microspheres are filled with butanol, the addition of particles leads to an enrichment. After the measurements, the fuel injection and ignition were disabled and the combustion engine was shut down, highlighted as (4) in Figure 3.14.

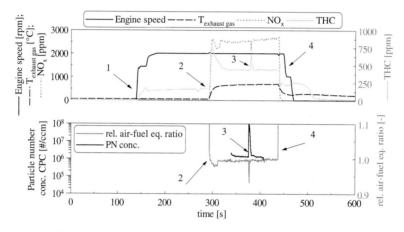

Figure 3.14: Effect of particle seeding on combustion characteristics and exhaust gas formation

3.5.3 Post-processing PIV

The high-speed camera provides an initial time series of raw images for each combustion cycle. To calculate the PIV vector fields, the time series of images was reorganised by the Software[4] to double-frame images for each consecutive time step (Δt), which is defined by the cfr of 24 kHz. The cross-correlation algorithm used was based on multi-pass iterations with two decreasing steps of interrogation window size (iws). For the first step, the initial window size was set to 24 x 24 pixels with rectangular shape and reduced to 6 x 6 pixels with a circular shape for the second step. The first step was calculated with two iterative passes and an overlap of 50 %, whereas the second step was calculated three times with an overlap of 25 %. The final multi-pass post-processing using a regional median filter to remove spurious vectors replaced all vectors with a deviation of 2.5 times the RMS of the neighbours. All vector groups with less than 5 vectors were thereby removed.

---

[4]DaVis 8.3 by LaVision

For the calculation of the flow velocities, several limitations exist. The maximum detectable flow velocity generally depends on the camera framing rate, the setting of the interrogation window size and shape as well as the velocity gradients of the tracer particles within the interrogation windows. Assuming that a tracer particle takes a displacement of 75 % of the final iws (75 % of 6 px, = 1.76 mm) and using the cfr of 24 kHz, a maximum velocity of 42.24 m/s is theoretically detectable. Concerning the minimum detectable velocities, the cross-correlation function needs to be analysed. Depending on the quality of the raw data, the position of the cross-correlation peak has a sub-pixel accuracy of $1/10^{th}$ to $1/20^{th}$ of a pixel, according to Raffel et al. [153]. For rather narrow correlation peaks, the Gaussian peak fit three-point estimator achieves the best results when the diameter of the particle image is slightly above 2 pixels. For the camera framing rate of 24 kHz, the smallest flow velocity is theoretically about 0.94 m/s. The theoretical detection limits are shown in Table 3.4. The theoretical measurement uncertainties of PIV were described by Keane and Adrian [98] as well as by Westerweel [200]. They describe the dependency of the uncertainty of the particle density. A higher number of particle image pairs in the observation area improves the signal strength of the cross-correlation peak and thus the valid detection limit [98]. The maximum possible particle image density is limited by the contrast of the PIV images.

Concerning the use of microspheres in a combustion engine, Disch et al. showed that an increase of the mean-averaged particle diameter of up to 250 µm was detectable in the late compression stroke [39]. They assumed that on the one hand, the increasing particle density in the late compression stroke supports the agglomeration of particles and on the other hand, the hollow thin-shelled microspheres diameter increased due to the increase in temperature during compression.

In addition to the flow-field calculation described beforehand, the analysis of the kinetic energy is helpful for the interpretation of the measured data. Therefore, the mass-based average kinetic energy (E_{AKE}, equation 3.5) and the mass-based average

Table 3.4: Theoretical limits for maximum and minimum detectable velocities.

Engine speed	[rpm]	2000
Camera frame rate	[kHz]	24
Δt	[s]	4.17e-5
interrogation window size (6 pixel)	[mm]	2.35
max. velocity (75% iws*)	[m/s]	42.4
min. velocity (0.1 pixel**)	[m/s]	0.94

* according to Keane and Adrian [98]
** according to Raffel et al. [153]

turbulent kinetic energy (E_{TKE}, equation 3.6) of the vector fields were calculated with the following equations.

$$E_{AKE} = \frac{1}{2} \cdot (v_{avg})^2 = \frac{1}{2} \cdot \left(\frac{1}{n} \cdot \sum_{i=1}^{n} v_i \right)^2 \tag{3.5}$$

$$E_{TKE} = \frac{1}{2} \cdot \left(\sqrt{\frac{1}{n-1} \cdot \sum_{i=1}^{n} (v_i - v_{avg})^2} \right)^2 \tag{3.6}$$

In these equations v_i is the velocity of vector i and n is the sum of all vectors taken into the calculation. As the light sheet was oriented in the central cross-section of the cylinder, only the two-dimensional in-plane vectors could be taken into account for the calculations. This is disadvantageous especially for the measurement of the flow-field at engine operation with a swirl motion. However, a change of the arrangement of the light-sheet and the camera system to detect the flow-field in the horizontal plane of the engine was not possible.

3.6 Pressure chamber

First investigations to show the influence of the reduced static flow of the injectors and the increased injection pressure respectively, were performed in a pressure chamber. The technical data of the pressure at the IFKM are shown in Table 3.5. The pressure chamber provides a better optical accessibility in comparison to a combustion engine, due to large glass windows, as shown in Figure 3.15. Furthermore, constant operating parameters, such as pressure and temperature, can be adjusted repeatably. The chamber has five identical accesses, which can be used either with a glass window for a camera system and the illumination, or as an adapter for the injector. To purge the chamber, a nitrogen volume flow can be adjusted appropriately. The heating of the chamber is realised via heating elements and heating sleeves.

Table 3.5: Technical data of the pressure chamber [174].

Max. pressure in the chamber	[MPa]	7.5
Cracking pressure of the safety valve	[MPa]	7.0
Working pressure	[MPa]	<7.0
Max. temperature	[°C]	450
Max. temperature using purging	[°C]	≈ 350
Volume of the chamber	[dm³]	≈ 6.2
Number of glass windows		5
Diameter of the glass windows	[mm]	95

For these investigations, one access was used for the injector, while the access on the opposite side was used for illumination and a CCD camera. The high-speed camera was mounted on the orthogonal access to the injector, as shown in Figure 3.15. For the illumination, two light sources were used: A cold-light source operating in permanent mode and a high power diode operating synchronised to the frequency of the high-speed camera at 24 kHz. The high power diode operates at 528 nm and is visible as green light. The drive electronics for the diode were manufactured at IFKM according to the recommendation of Willert et al. [204].

The injector was oriented in two different positions for the measurements: The first orientation was comparable to the engine. Thereby the high-speed camera was in the viewing angle of the spark plug. To get a second view on the spray, the injector was rotated by about 90°.

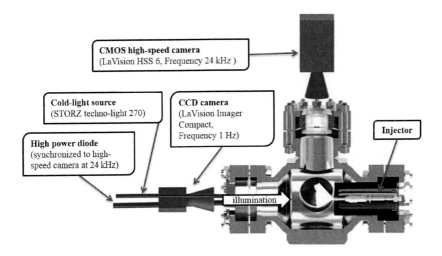

Figure 3.15: Setup of the pressure chamber [108]

4 Pressure chamber results

The investigations to reduce particle emissions of GDI engines is separated in two parts. The first part in chapter 4 summarises the characterisation of the injectors used in the project. The injectors were operated in a pressurised chamber under different conditions. The injection pressure as well as the pressure and the temperature in the pressurised chamber were varied consistent with the later on following engine operation (homogeneous, pre-mixed operation at high engine load). The second part in chapter 5 describes the results of the investigations using a single cylinder research engine. The engine was operated under steady state conditions and different parameters were individually varied to show their influence on the particle formation process.

4.1 Injector characterisation in the pressure chamber

To compare the two injectors used, results of investigations in an optically accessible pressure chamber are shown in Table 4.1. A pressurised chamber is used to individually investigate different influences on the injection behaviour and the spray formation of the two injectors used. The boundary conditions such as the temperature and the pressure in the chamber can be chosen independently. In a combustion engine, the movement of the piston as well as the charge air motion influence the spray formation. The walls of the cylinder liner also restrict the spray penetration. In a pressurised chamber there is no air motion and the volume of the chamber is large enough not to have any interaction of the spray with the walls. Additionally in a combustion engine the pressure and the temperature are not constant during the injection process. Therefore, the results of the pressurised chamber can show differences to investigations in a combustion engine. However, differentiated investigations of influencing factors on the spray formation with a good optical accessibility are not possible in this detail in a combustion engine.

The investigations to compare the two injectors used in this project were performed under conditions comparable to the boosted operation of the single cylinder engine with a coolant temperature of 90 °C and a pressure of $0.12 \, \mathrm{MPa_{absolute}}$ in the chamber. The six-hole injector was mounted such that the high-speed camera was in the viewing angle of the spark plug in the engine[1]. The pictures were taken with a framerate of 24 kHz. Mie-Scattering images at different time steps after start of injection (SOI) are

---

[1]Injector is mounted in plane of the crankshaft. Two of the six injection jets are oriented towards the spark plug

shown in Table 4.1 on page 65. A threshold value was set to 150 counts to reduce the background noise. 30 injections were averaged for each setting.

The injection pressure was varied from 20 MPa to 35 MPa for the high flow injector and for the low flow injector additionally to 50 MPa. The shorter delay time between electrical and hydraulic start of injection is obvious for the high flow injector. Thus at earlier timings, for instance at 0.46 ms after SOI, a noticeable amount of fuel is injected using the high flow injector. For the low flow injector the delay between the electric and hydraulic SOI is longer and thus only a small amount of fuel is injected at 0.46 ms after SOI. Caused by the higher hydraulic flow, the penetration velocity of the high flow injector is higher compared to the low flow injector. The intensity information of each pixel is displayed in the colour information. A higher intensity is represented by a deeper colour. It is obvious that for higher injection pressure, the intensities of scattered light are higher for both injectors. According to Hoffmann et al. [84], the higher scattered intensity is an indicator of the droplet-size reduction associated with the system pressure increase. With the smaller droplet size the droplet surface increases and more light is scattered. Another important information of the images is the comparable spray angle for the two injectors. As the spray targeting of the two injectors was similarly designed, the comparable spray angle shows that the two injectors fulfil the requirements.

4.2 Spray penetration and break-up as a function of hydraulic flow and injection pressure

In Figure 4.1, the calculated spray penetration length of the measurements in the pressure chamber is visualised. The high flow injector was operated with 20 MPa and 35 MPa injection pressure and the low flow injector with 20 MPa, 35 MPa and 50 MPa respectively. The energising time was set for the same injected fuel mass. As the same fuel was used, the end of injection depends on the static flow and the injection pressure. The calculation was realised from electrical SOI (set to 0 ms) to hydraulic end of injection (EOI). It is to mention that the maximum measurable penetration length in the pressure chamber used was 80 mm, due to the dimensions of the observation window. In the data loupe in the corner on the lower right hand side, the hydraulic start of injection is zoomed in. It is obvious, that the high flow injector has a shorter time between electrical and hydraulic SOI (SOI delay). For the high flow injector, the SOI delay is about 0.33 ms while the SOI delay for the low flow injector is about 0.42 ms. At 2000 rpm this time corresponds to 3.6 °CA for the high flow injector and 4.8 °CA for the low flow injector respectively. With increased injection pressure, the penetration length rises for the same time after SOI for both injectors. The maximum penetration length at hydraulic end of injection is thereby on a similar level for both injectors. Using the same injection pressure, the injection duration is longer for the low flow injector because of the reduced static flow. However, the maximum penetration length is higher for the low flow injector using the same injected fuel mass. This indicates that at engine operation, a fuel-piston-interaction is more likely using the low flow injector at the same engine load. With increased injection pressure, the higher penetration length is reached at earlier timings.

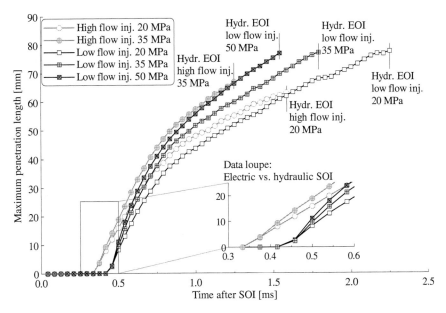

Figure 4.1: Calculated spray penetration length of the pressure chamber measurements at $T_{chamber} = 90\,°C$, $p_{chamber} = 0.12\,MPa$, equal injected fuel mass for the different injection pressures and hydraulic flows

4.3 Initial penetration velocity as a function of hydraulic flow and injection pressure

The initial penetration velocity, Figure 4.2, also shows the shorter SOI delay of the high flow injector. The maximum penetration velocity is higher for the low flow injector. Also the maximum penetration velocity depends strongly on the injection pressure. With increased injection pressure the penetration velocity is increased significantly. The highest penetration velocity is measured shortly after the hydraulic start of injection. This first phase of the injection is called the primary spray break-up. Thereby the fuel jet breaks up in small droplets and ligaments close to the injector tip. The injection pressure, nozzle geometry as well as the physical properties of fuel and the surrounding medium influence the primary break up significantly. After the primary break-up, the penetration velocity decreases and approaches a constant penetration velocity of 15 m/s to 30 m/s. This second part of the injection process is called the secondary break-up and is caused by the relative velocity between fuel droplet and gas phase. The surface tension of the fuel droplets acts against the aerodynamic forces caused by the interaction of the fuel droplets and the surrounding air [117]. The behaviour of the droplets is defined by the deceleration due to aerodynamic drag, convective heat transfer from the entrained air, and mass transfer of vaporised fuel away from the droplet [8]. The maximum velocities measured are consistent to the measurements by Leick [118]. Leick measured the spray velocimetry of Diesel sprays in a pressurised chamber with different injection pressures (22 MPa and 80 MPa) and pressures in the chamber (0.1 MPa and 2.0 MPa) using laser correlation velocimetry.

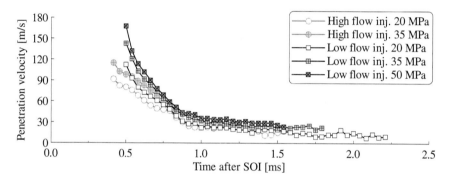

Figure 4.2: Calculated spray penetration velocity of the pressure chamber measurements at $T_{chamber} = 90\,°C$, $p_{chamber} = 0.12\,MPa$, equal injected fuel mass for the different injection pressures and hydraulic flows

As shown in the previous graphs, the injection is divided in two parts: the primary break-up and the secondary break-up of fuel. Caused by the high penetration velocity in the primary phase, the spray area visible in the high-speed images increases fast in this phase. During the following secondary break-up, the spray area further increases until the end of injection. The calculated spray area is shown in Figure 4.3. Using the high flow injector, the injection starts earlier because of the shorter SOI delay. Therefore, more fuel is injected into the pressurised chamber at early timings. However the spray area using the high flow injector and the low flow injector with 20 MPa is on the same level until the end of injection of the high flow injector.

For the same injection pressure, the low flow injector delivers a smaller amount of fuel using the same injection timing. Thus the same spray area can be explained by the smaller injected fuel droplets, as shown in Section 3.2.2, and thus stronger entrainment of air into the spray. For the increased injection pressure of 35 MPa, the same behaviour is obvious. Even though the injection using the high flow injector starts earlier, the spray area is on a comparable level using the low flow injector. Besides the hydraulic flow of the injector, the injection pressure influences the spray area significantly. With increased injection pressure, the SMD decreases and the kinetic energy of the fuel droplets increases. Thus the spray penetration length as well as the spray area increase. With the reduced SMD the evaporation process is enhanced because the surface to volume ratio increases with smaller droplet sizes. After the hydraulic end of injection, the spray area decreases because no more fuel is injected and the fuel in the chamber evaporates. For the higher injection pressure, the evaporation process is faster and thus the spray area decreases faster for both injectors.

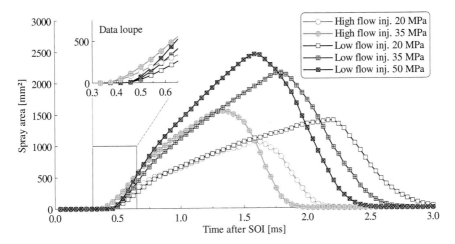

Figure 4.3: Calculated spray area of the pressure chamber measurements at $T_{chamber} = 90\,°C$, $p_{chamber} = 0.12\,MPa$, equal injected fuel mass for the different injection pressures and hydraulic flows

4.4 Changes of temperature and pressure in the chamber

The previously shown results were all performed at constant boundary conditions in the pressure chamber (pressure in the chamber $= 0.12\,MPa$; temperature in the chamber $= 90\,°C$). As the boundary conditions have a significant impact on the injection characteristics, the influence of the pressure and the temperature in the chamber were investigated. Therefore, both injectors were operated with the different injection pressures (20 MPa, 35 MPa and 50 MPa respectively) and both temperature (25 °C, 90 °C and 120 °C) and pressure (0.12 MPa and 0.18 MPa) were varied independently. These values of temperature and pressure were chosen based on the conditions in the combustion engine for homogeneous injection at boosted operation. The calculated penetration length at hydraulic end of injection is plotted versus the relative SMD, as defined in section 3.2.2, for the different temperatures of the chamber in Figure 4.4. In the upper graph the penetration length are shown for a pressure of 0.12 MPa and for 0.18 MPa in the lower graph. The penetration length increases with higher injection pressure for both injectors. The temperature of the chamber shows significant influence on the penetration length. With increased temperature, the evaporation process is enhanced and thus the penetration length decreases. With an increased pressure in the chamber, the penetration length slightly decreases. This is caused by the increased density in the chamber [118].

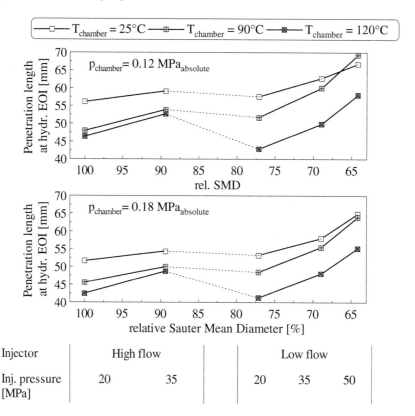

Injector	High flow		Low flow		
Inj. pressure [MPa]	20	35	20	35	50

Figure 4.4: Influence of temperature and pressure in the chamber on penetration length for different injection pressure and hydraulic flow with equal injected fuel mass for the different injection pressures and hydraulic flows

Table 4.1: Spray pattern of low flow injector and high flow injector for different injection pressures. Injector specifications shown in table 3.3

Time after SOI	High flow inj. 20 MPa	High flow inj. 35 MPa	Low flow inj. 20 MPa	Low flow inj. 35 MPa	Low flow inj. 50 MPa
0.46 ms					
0.58 ms					
0.79 ms					
1.00 ms					
1.42 ms					
1.88 ms					

pressure in the chamber	$= 0.12\,\mathrm{MPa_{absolute}}$
temperature in the chamber	$= 90\,°\mathrm{C}$
averaged number of pictures	$= 30$ for each setting

0   0.2   0.4   0.6   0.8   1

Normalized intensity of the scattered light [-]

5 Engine results

Different influencing factors on particle number emissions were investigated in the project and are discussed in this chapter. The first section focuses on the influence of the charge motion on the mixture formation process. Thereby different strategies to influence the in-cylinder charge motion were investigated. Besides the variation of the intake valve timings, the implementation of a large scale charge motion by selectively closing parts of the intake channel showed promising results for decreasing particle number emissions. By reducing the maximum intake valve lift of the mechanical variable valve train, an early intake closing can be realised. Due to the charge air pressure needing to be increased in order to compensate the shorter intake open timing, an increased flow velocities at the intake valve are assumed during intake open timing (ivo).

The next part of the chapter focuses on changes of the gas mixture. Firstly the air-fuel equivalence ratio (λ) was varied. Modern gasoline engines operate under rich conditions at high engine load to keep the engine from damages. This enrichment is supposed to increase the particle number concentration because of the combustion under fuel rich conditions.

Another influencing factor both on the caloric and the peak temperature in the cylinder is the amount of residual gas. The amount of residual gas depends on the charge exchange process and is influenced by the gas dynamics as well as the flow ratio and the pressure ratio. To vary the residual gas amount, the valve overlap and the exhaust back-pressure were varied.

Another way to change both the caloric and the peak temperatures in the combustion chamber is to implement an exhaust gas recirculation (EGR) system. In this project, a high pressure EGR system was installed in combination with a heat exchanger.

Besides the charge motion inside the cylinder, the injection system is the main influencing factor on the mixture formation process. The injector characteristics such as the hydraulic flow ($\dot{Q}_{stat}$) and the injection pressure influence the Sauter Mean Diameter (SMD). In this project, two injectors with the same injection targeting but different hydraulic flows were used with different injection pressures. Besides the nozzle characteristics, the timing of the injection also determines the quality of the mixture formation process.

Due to the great importance of the fuel composition and conditions (temperature and pressure), investigations were performed with increased injection pressure and increased fuel temperature. Alkylate fuel was additionally used with comparable physical properties to RON 95, whereas the chemical composition is different. Therefore, the influence of the polycyclic hydrocarbons on the soot formation can be separated.

Independent variations of the coolant and lubricant temperature were also performed, showing the influence of lower piston temperature caused by reduced lubricant temperature and reduced cylinder head and liner temperatures for the reduction of the coolant temperature.

The last influencing parameter on PN formation in GDI engines investigated in this work is the inflammation. For the above mentioned parameter variations, a state-of-the-art transistor coil ignition was used. To enhance the inflammation process, a high-frequency ignition system was tested.

5.1 Influence of large scale charge motion

Different strategies to generate a large scale charge motion were investigated during the project. Three inlays were used to generate a tumble flow as well as one inlay to generate a swirl motion, as shown in Figure 3.3. To show the influence of the charge motion as well as the impact of different valve timings, a variation of the intake valve open timing was performed. A further increase of the turbulent kinetic energy can be realised by early intake valve closing. Therefore, the maximum valve lift of the single cylinder research engine was reduced and a variation of the intake valve open timing was realised. The ignition timing was set to knock-limited spark advance[1]. The injection timing of the high flow injector was fixed for all settings at $SOI = -280\,°CA_{\,aTDCf}$ whereas the injection pressure was kept constant at $20\,MPa$. The exhaust valve timings were fixed, therefore the variation of the intake open timing is accompanied by a change of the valve overlap. The exhaust back-pressure was set according to a reference engine at the reference point and changed $1:1$ with the charge air pressure. An extract of these results was published at SIA conference 2015 in Versailles [14].

5.1.1 Large scale charge motion using inlays

The first tests were performed using different charge motion inlays with full intake valve lift at $2000\,rpm$ and $1.4\,MPa$ IMEP and the high flow injector. The exhaust valve spread was fixed for exhaust valve closing close to TDC. The resulting particle number concentrations of the intake open timing variation are shown in Figure 5.1. Using the baseline configuration, the particle number emissions are at the highest level. The particle number emissions can be reduced by using the charge motion inlays. Lowest particle emissions can be achieved by using the swirl configuration. Using the tumble inlays, the 50 % closed channel shows the best performance compared to the 60 % closed version and the 70 % closed version.

---

[1]The knock limit was defined in a way that less than 3 % of the measured cycles showed a knocking amplitude of $0.1\,MPa$ or higher.

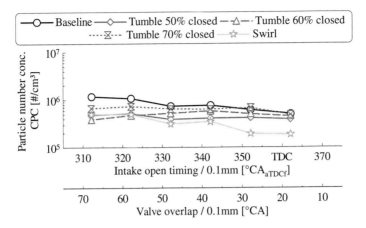

Figure 5.1: Influence of large scale charge motion on PN emissions at 2000 rpm, 1.4 MPa IMEP, SOI: -280 °CA, $p_{Inj.}$: 20 MPa, high flow injector, ignition timing: knock-limited spark advance

The particle number concentration can also be reduced by applying a later intake open timing. With the later intake open timing the valve overlap is reduced as well as the effective compression ratio as shown in Figure 3.4. To understand the reduction of the PN concentration with later intake open timing, additional results of the exhaust gas analysis and the thermodynamic analysis of the variation are shown in Figure 5.2.

Using the large scale charge motion, the CoV of IMEP can be reduced. This is mainly caused by a better mixture formation and a shorter burning duration. The MFB50 % can be shifted to earlier timings because of the lower knocking tendency for all charge motion strategies except for the 70 % closed tumble inlay. It is assumed that the inlay with the 70 % closed intake channel does not form a fully developed tumble. Additional measurements using PIV were done to validate this assumption and are discussed in the next section. The residual gas amount is increased with later intake valve open timing as a result of reduced valve overlap. For the earlier intake open timings and the linked larger valve overlap, scavenging is possible. The increasing amount of carbon monoxide in the exhaust gas can be explained by the incomplete combustion resulting from the slightly rich conditions in the combustion chamber caused by scavenging. It is assumed that the reduction of the particle number concentration at later intake open timings has two reasons: firstly, the lower scavenging effect and secondly the higher temperatures at injection caused by the higher amount of residual gas and the resulting enhanced mixture preparation. These findings agree well with the investigations of Piock et al. who found reduced particle emissions with enhanced in-cylinder charge motion [147]. Piock et al. thereby used two levels of tumble charge motion in comparison to a configuration with low charge motion and state that the effect of tumble charge motion enhancement is the promotion of mixture homogeneity and thus reduced particle emissions. To quantify the in-cylinder charge motion, particle image velocimetry (PIV) measurements were performed. The setup of the system and the test procedure is shown in section 3.5. To minimize the influence of scavenging, the latest intake open timing, resulting in the lowest valve overlap, of the previous tests was chosen for the measurements. The results of the flow investigations are shown in Figure 5.3 as numerical analysis and in Figure 5.4 as flow visualisation in terms of streamlines[2].

In the upper graph of Figure 5.3 the average velocity is shown versus the crank angle, while the turbulent kinetic energy is shown in the bottom graph. On the y-axis on the right hand side of each graph, the intake valve lift is plotted. The grey boxes in the graphs represent the injection time. As the injection of the fuel was directed into the calculation area of the light sheet and spatially set between light sheet and high speed camera, the scattered light of the particles in the light sheet is blocked by the injected fuel spray. Therefore, the calculated velocities during injection are not representative for the flow field. After the end of injection, the influence of the charge motion configuration increases significantly. As assumed beforehand, the average flow velocities after injection are the highest using the swirl charge motion, resulting in enhanced mixture formation and a rapid inflammation and combustion process. During intake

---

[2]A streamline is defined as line that is at any position parallel to the local velocity vector

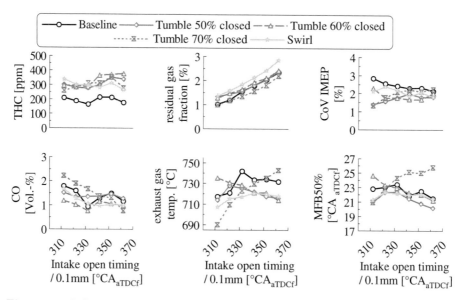

Figure 5.2: Influence of large scale charge motion on combustion characteristics at 2000 rpm, 1.4 MPa IMEP, SOI: -280 °CA, $p_{Inj.}$: 20 MPa, high flow injector, ignition timing: knock-limited spark advance

open timing after injection, the measured flow velocities of the baseline case and the tumble cases 50 % closed and 60 % closed show comparable flow velocities, whereas the tumble configuration with 70 % closed intake channel shows significantly lower flow velocities. The reason for these findings is shown at the streamline visualisations in Figure 5.4. The baseline case and the tumble cases 50 % closed and 60 % closed show a directed flow from intake valve to the spark plug direction. For the 70 % closed case, a diffuse flow field was measured, indicating that the closing of 70 % of the lower part of the intake channel did not lead to a generation of a tumble motion inside the cylinder. These findings give an explanation for the increased PN concentrations measured and the engine being more prone to knocking with this charge motion configuration. During the compression stroke, after intake valve closing, the flow velocities increase due to the piston motion in upwards direction. For the baseline configuration, a homogeneous flow from centre to the top of the combustion chamber is obvious at -80 °CA $_{\mathrm{aTDCf}}$ in Figure 5.4. Using the tumble configurations 50 % and 60 % closed, a tumble motion was measured, which increases the flow velocities additionally. Comparing the flow velocities of the 70 % closed tumble inlay and the swirl configuration, the same average velocities were measured at the end of the compression stroke. However, as the laser light sheet was positioned parallel to the swirl axis, the main flow of the swirl is assumed to flow through the measurement plane and thus can not be measured, as the PIV technique is only able to measure the two-dimensional velocities. Therefore, it is assumed that the flow velocities and turbulent kinetic energy using the swirl configuration is higher than the measured velocities in the layer chosen for PIV.

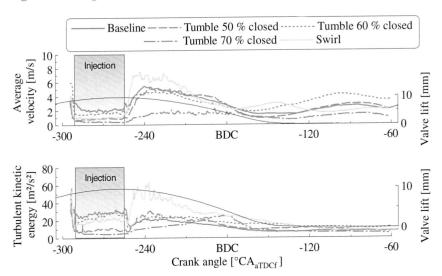

Figure 5.3: Results of the flow investigations using high speed PIV at 2000 rpm, 1.4 MPa IMEP, SOI: -280 °CA, p$_{\mathrm{Inj.}}$: 20 MPa, high flow injector, ignition timing: knock-limited spark advance, ivo for minimum valve overlap

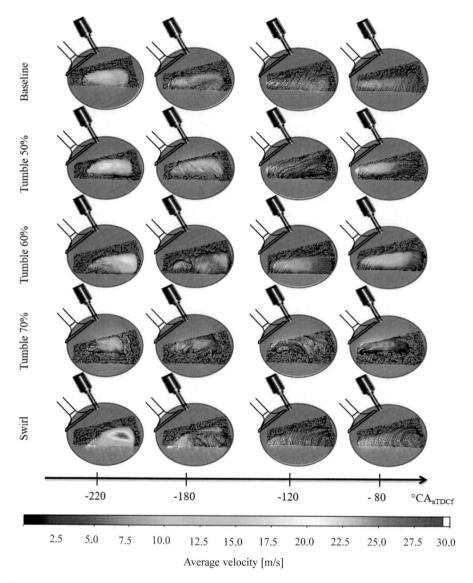

Figure 5.4: Streamline visualisation of the flow field for the different charge motion strategies at 2000 rpm, 1.4 MPa IMEP, SOI: -280 °CA, $p_{Inj.}$: 20 MPa, high flow injector, ignition timing: knock-limited spark advance, ivo for minimum valve overlap

5.1.2 Large scale charge motion and early intake closing

The next strategy to enhance the mixture preparation investigated was early intake closing. To realise an early intake closing using the mechanical valvetrain of the engine, the intake valve lift needs to be reduced. Three different maximum valve lifts were investigated: 5 mm, 6 mm and 7 mm. Thereby the valve spread was adjusted to the same intake valve open timings as in the previous section. The emitted particle number concentrations for the variations are plotted versus the intake open timing in Figure 5.5 with identical scaling as shown in Figure 5.1.
For these investigations the tumble inlay with the 60 % closed intake channel of the previous section was used as the baseline for the comparison. Because of the range of the phase shifters, not all intake open timings were possible for the lower intake valve lifts. Nevertheless a further reduction in particle number concentrations is possible using early intake closing. The only exception is the earliest intake open timing using the 7 mm maximum intake valve lift. The lowest PN concentration is achieved by reducing the maximum intake valve lift to 5 mm.

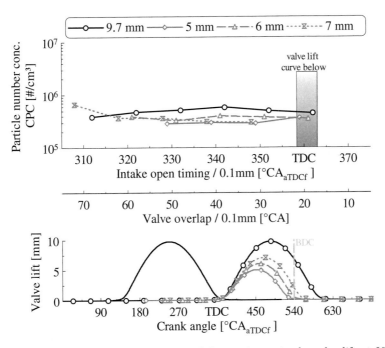

Figure 5.5: PN emissions for the variation of the maximum intake valve lift at 2000 rpm, 1.4 MPa IMEP, SOI: -280 °CA, $p_{Inj.}$: 20 MPa, high flow injector, ignition timing: knock-limited spark advance

Additional thermodynamic data and the exhaust gas analysis for these investigations are shown in Figure 5.6. Caused by the early closing of the intake valve, the boost pressure needs to be increased to compensate the shorter intake open timing. For early intake open timings, the intake valve closes before bottom dead centre resulting in an expansion of the cylinder charge. Thus the temperature in the combustion chamber is decreased. In addition, scavenging increases with earlier intake open timings due to the increasing valve overlap. These two effects are supposed to be responsible for the higher CO emissions and earlier MFB50 % at early intake closing. Scavenging is also assumed to cause the high emission of PN using the 7 mm intake valve lift and the earliest intake open timing. The CO emission is on the highest level for all of the variations within this configuration. Overall the investigations with the 5 mm maximum intake valve lift show the lowest emissions considering particle number concentration and gaseous exhaust emissions. However, the required boost pressure is on the highest level.

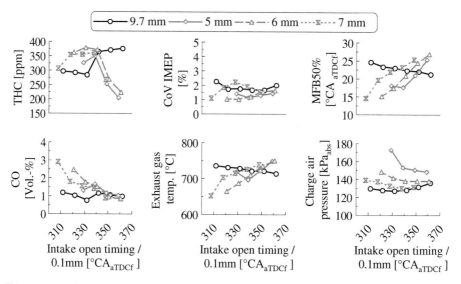

Figure 5.6: Combustion characteristics for the variation of the maximum intake valve lift at 2000 rpm, 1.4 MPa IMEP, SOI: -280 °CA, $p_{Inj.}$: 20 MPa, high flow injector, ignition timing: knock-limited spark advance

To estimate the influence of the different valve timings and valve lifts, the TPA model was used. The resulting turbulent kinetic energy at start of injection as well as the temperature at ignition timing for the investigations with early intake closing are shown in Figure 5.7. The effect of the early intake closing on the temperature at ignition timing is shown in the graph on the right hand side. Once again, the 5 mm maximum valve lift shows the highest potential to reduce the in-cylinder temperatures. Due to the lower in-cylinder temperatures, the engine is less prone to knocking and can be operated at advanced combustion phasing, resulting in higher efficiency. Thereby the turbulent kinetic energy at start of injection is also increased resulting in a better mixture formation and therefore in lowest particle number concentrations, as shown in Figure 5.5.

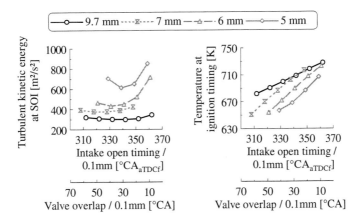

Figure 5.7: Calculated turbulent kinetic energy at SOI and temperature at ignition timing (TPA) for the variation of the maximum intake valve lift at 2000 rpm, 1.4 MPa IMEP, SOI: -280 °CA, $p_{Inj.}$: 20 MPa, high flow injector, ignition timing: knock-limited spark advance

5.2 Variation of air-fuel-ratio

Most gasoline engines operate with a stoichiometric air-fuel ratio ($\lambda = 1$) within most of the engine map. At high engine load however, the engine manufacturers apply the engines with enriched mixture ($\lambda < 1$). The additional fuel is used to lower the in-cylinder gas temperatures using the evaporation heat. Furthermore, due to the reduced in-cylinder peak temperatures caused by the rich mixture, the exhaust gas temperature is decreased and saves the components of the exhaust part (exhaust valves, turbocharger, catalyst, etc.) from damage. However, operating the engine with a rich mixture leads to a combustion under oxygen deficiency. The process of soot formation depends to a large extent on the availability of oxygen and the temperature [20]. When operating the engine with an early injection timing during the intake stroke, it is assumed that the mixture is formed properly and the amount of fuel-rich zones in the gas phase should be low.

Sabathil et al. showed that for low engine load operation, the PN emissions can be increased in excess of factor 6 as the air-fuel ratio is shifted from $\lambda = 1$ to $\lambda = 0.85$. To show the effect of a rich mixture at boosted operation, the single cylinder was operated at 2000 rpm and the charge air pressure was adjusted to keep the engine load constant at 1.4 MPa IMEP. The injection timing of the high flow injector was chosen at minimum PN emissions for stoichiometric operation at 290 °CA $_{aTDCf}$ and 20 MPa injection pressure and kept constant for the variation. The engine was originally designed for natural aspiration and thus could not be operated at thermodynamic optimised MFB50% due to engine knocking. The self-ignition causing the knocking sound of the engine depends on the local air-fuel ratio and the local temperature in the unburned gas. Rothe showed, that local fuel-rich zones are the reason for temperature inhomogeneities in the end gas and that engine knocking starts in end gas zones with locally raised temperatures [162]. Thus the flame speed as well as the temperatures inside the cylinder influence the knocking behaviour. To keep the MFB50% constant for the whole variation, the ignition timing was set to a late centre of combustion (MFB50% = 28 °CA $_{aTDCf}$). The valve timings were set to a minimum valve overlap and thus reduced the scavenging effect. The engine back-pressure was set to the reference engine again and changed 1 : 1 when changing the charge air pressure.

The measured particle number concentrations are shown in Figure 5.8. On the left hand side are the particle number concentrations measured by the particle counter (CPC) plotted versus the air-fuel ratio (λ). While on the right hand side the summed up particle number concentration measured by the EEPS are plotted. Even though the EEPS measures particles with an electrical mobility diameter below 23 nm, only the particles with a bigger mobility diameter were considered to ensure a comparability of the particle measurement devices. A difference between the measured particle number concentrations can be noticed. This difference can be referred to the different measurement principle of the two systems. Nevertheless the tendencies of the measurements are comparable while the absolute value is slightly different. A detailed comparison of the two measurement systems CPC and EEPS is given by Khalek et al. [100].

For a wide range of air-fuel ratio, the PN emissions are on a constant level ($0.9 < \lambda < 1.0$). Further reducing the air-fuel ratio increases the PN emissions using the swirl inlay. Using the tumble inlay the PN emissions stay just about constant even at further enrichment. For mixtures leaner than stoichiometry, the PN emissions can be reduced. This is assumed to be caused by the dilution of the air on the one hand and by the excess supply of oxygen on the other. Therefore, the probability of fuel-rich zones is less likely and a post oxidation is possible.

Exhaust gas measurements and the combustion characteristics are shown in Figure 5.9. In general, the CO- and THC-emissions are directly depending on the air-fuel ratio and decrease with a leaner mixture. Incomplete combustion products like hydrocarbon and carbon monoxide are an indicator for combustion efficiency and increase with oxygen deficiency [79]. The carbon monoxide emissions are on the same level for both charge motion strategies (tumble and swirl), while the hydrocarbon emissions are higher using the swirl inlay. The formation of nitrogen oxides depends strongly on the temperatures during combustion and the oxygen availability [115]. The nitrogen oxide emissions therefore are the highest for $\lambda = 1.1$ and strongly reduced for the richer mixtures. The exhaust gas temperature decreases both for lean and rich mixtures

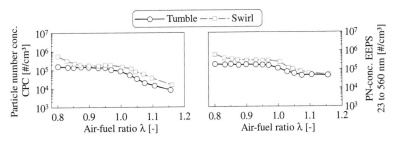

Figure 5.8: PN concentration for the variation of air-fuel ratio λ at 2000 rpm, 1.4 MPa IMEP, SOI: -290 °CA, $p_{Inj.}$: 20 MPa, high flow injector, ignition timing: knock-limited spark advance

while the peak temperature is at slightly lean mixtures. These effects are caused by insufficient oxygen concentration and the cooling effect of the evaporating fuel for rich mixtures and the charge dilution for leaner mixtures respectively.

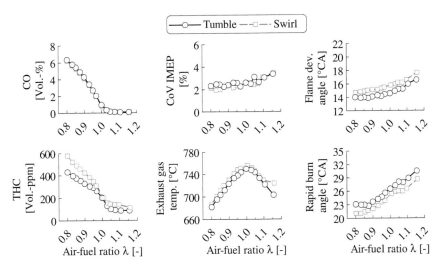

Figure 5.9: Combustion characteristics for the variation of air-fuel ratio λ at 2000 rpm, 1.4 MPa IMEP, SOI: -290 °CA, $p_{Inj.}$: 20 MPa, high flow injector, ignition timing: knock-limited spark advance

At this high engine load and late centre of combustion (MFB50%), the cyclic variations are strong. Because the combustion starts at later crank angle degree, the turbulence is on a lower level at ignition and during the combustion, thus leading to a less robust inflammation and slower combustion [79]. This leads to an increased flame development angle[3] as well as a prolonged duration of combustion (rapid burn angle[4]), accompanied by stronger fluctuations and thus increased cyclic variations (CoV IMEP). To get more information on the causes of the PN emissions for the variation of the air-fuel ratio, measurements with the high-speed camera were performed. The setup of the high-speed camera is shown in section 3.4.3. For these investigations, no additional illumination was used to visualise the fuel injection (see section 3.4.1). Therefore, only the soot luminescence is visible on the high-speed images, as described in section 3.4.2. For the post-processing of the high-speed measurements, the intensity information of each pixel, the average intensities as well as the standard deviation of the intensities of each high-speed image were calculated. The standard deviation of the intensity is more sensitive to bright spots, which are a sign of diffusive combustion. The standard deviation of intensity is plotted for the 80 measured cycles versus the crank angle in Figure 5.10. In the upper, right hand side of each graph, a data loupe is placed for the late timing after TDC (55-65 $°CA_{aTDCf}$). This timing is relevant regarding soot emissions, because at this time of the cycle the combustion is nearly finished, but the temperature in the combustion chamber is high enough for the soot to emit light. In the upper three graphs, the three air-fuel ratios $\lambda = 1.1$, $\lambda = 1.0$ and $\lambda = 0.9$ using the tumble configuration are shown. In the lower graph the same λ values were used, however these graphs show the data using the swirl configuration.

The obvious cycle-to-cycle variations are typical for gasoline engines. These cycle-to-cycle variations are caused by variations in the charge motion within the cylinder at ignition timing, variations in the amounts of air and fuel fed into the cylinder as well as the mixing of the fresh air and the residual gas [79]. Noticeable are the lower maximum values of the standard deviation for the rich mixture. This is supposed to be caused by the lower maximum temperatures in the cylinder at rich operation, caused by the raised in-cylinder cooling due to direct injection of the higher fuel mass. The lower temperatures result from the increased evaporation heat needed to evaporate the higher fuel mass. Due to the higher laminar flame velocity, the combustion process is accelerated compared to the stoichiometric and lean operation, as already shown in Figure 5.9. For the lean case, the standard deviation is the highest for late timings, because of the longer burning duration. However, for the lean and the stoichiometric case, nearly all cycles are in a small scatter band after 40 $°CA_{aTDCf}$. The standard deviations are on a lower level for the fuel-rich cases, although some single cycles have higher values compared to the average values. To visualise this, the cycles with the exceptionally high standard deviation at 60 $°CA_{aTDCf}$ are highlighted in Figure 5.10. For these cycles, the high-speed images at 60 $°CA_{aTDCf}$ are shown in Figure 5.11. One exception is the stoichiometric swirl case, in which one cycle with a long flame development angle and following late centre of combustion was measured (dark grey, dashed in Figure 5.10). This cycle has the highest value at 60 $°CA_{aTDCf}$, which is not caused by diffusive combustion but rather the late burning. Another cycle of this case

---

[3]flame development angle FDA; time between ignition timing and start of combustion (MFB05%)
[4]rapid burn angle RBA; time between start of combustion (MFB05%) and end of combustion (MFB90%)

however, was measured with high standard deviation of intensity in the later phase of the cycle (black, solid).

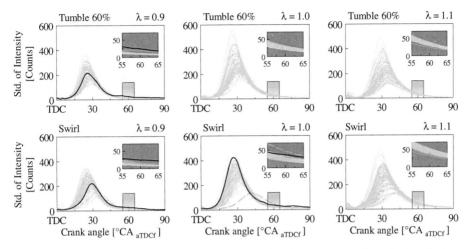

Figure 5.10: Analysis of the high-speed images for the variation of air-fuel ratio λ at 2000 rpm, 1.4 MPa IMEP, SOI: -290 °CA, $p_{Inj.}$: 20 MPa, high flow injector, ignition timing: knock-limited spark advance

For the stoichiometric and lean tumble case as well as the lean swirl case, no diffusive combustion was visible, which is why none of these cycles are shown. For the other three cases, the stoichiometric swirl as well as the fuel-rich with both charge motion strategies, a diffusive flame is visible in the gas phase. To illustrate the motion of this diffusive flame, four pictures are shown for each cycle with the timings: 50 °CA $_{aTDCf}$, 55 °CA $_{aTDCf}$, 60 °CA $_{aTDCf}$ and 65 °CA $_{aTDCf}$. In the upper row, the cycle of the rich tumble configuration highlighted in Figure 5.10, is shown. A diffusive flame is visible in the gas phase on the intake side. The diffusive flame being on a lower position for every 5 °CA indicates that the diffusive flame occurred in the gas phase and followed the piston movement downwards. This indicates that a fuel-rich zone existed in the gas phase, because of an improper mixture formation process and the overall rich air-fuel ratio.

Only one cycle of the stoichiometric swirl case was measured with a diffusive combustion after the regular combustion. In this cycle, the diffusive combustion was located in the upper right hand side part of the combustion chamber. The exhaust valves are located in this area. As the flame does not move throughout the expansion stroke, it is assumed that the diffusive flame was bound to the exhaust valve gap.

For the swirl case with a rich mixture in the lower row, the diffusive combustion is again on the left hand side of the images and moves downwards with the piston movement. It is again assumed, that this diffusive flame was caused by inhomogeneities in the mixture.

As shown in Figure 5.10, the cycle for the rich tumble and swirl case were not the only ones with diffusive combustion, indicating that the mixture formation process was influenced by the rich mixture in some cycles. However, not all cycles show a diffusive combustion, which is why PN emissions did not rise significantly through a slight enrichment.

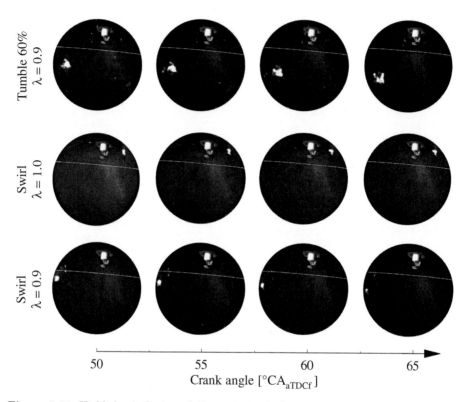

Figure 5.11: Highlighted Cycles of Figure 5.10 of the variation of air-fuel ratio λ at 2000 rpm, 1.4 MPa IMEP, SOI: -290 °CA, $p_{Inj.}$: 20 MPa, high flow injector, ignition timing: knock-limited spark advance

These results fit well to the findings of de Francqueville and Pilla, even though they investigated the effect of air-fuel ratio at catalyst heating operation [50]. For a variation of the enrichment of the air-fuel ratio, they state that the PN emissions increase due to increased injection duration and increased fuel mass injected. Additionally, they state that the lack of oxygen inhibits the soot oxidation.

Sabathil et al. also investigated the influence of the air-fuel ratio at low engine speed (1000 rpm) and low engine load (≈ 0.4 MPa IMEP)[164]. They show that a significant increase in PN emissions was measurable for global air-fuel ratios $\lambda \leq 0.9$.

Jiao and Reitz attempted to model ratio effects on particulate formation in an SI engine under premixed conditions with a CFD model [88]. The calculations are based on experimental data by Hagemann and Rothamer using a PFI engine [69]. Jiao and Reitz state:

"Before the flame front reaches the cylinder walls, soot is mainly produced within the flame region, and soot residing in the burnt regions is reduced with sufficient residence time at high temperatures by oxidation. A higher mass fraction of soot is left in the centre of the cylinder under rich combustion condition.

Soot oxidation is favoured when: (1) the flame front reaches the cylinder walls. (2) No more species for soot inception and surface growth are left after the combustion.

Strong coagulation is observed in the centre of the cylinder when the local temperature is high enough, leading to a lower density and larger particles. Those large particles are simultaneously oxidised due to high OH concentrations in high temperature regions.

For both lean and rich combustion conditions, high soot levels were observed near the cylinder walls. Small particles were found in the centre of the cylinder due to the longer residence time for soot oxidation as the flame front propagates away. Large particles tend to be located near the cylinder walls due to surface growth from the species produced from the flame front."

5.3 Influence of residual gas concentration

The amount of residual gas inside the cylinder during compression is determined by the charge exchange process. Directly influenced by the residual gas fraction is the volumetric efficiency and the performance of the engine. Whereas the efficiency and the exhaust gas emissions are influenced by the effect of the residual gas on the working-fluid thermodynamic properties. The residual gas fraction is primarily a function of inlet and exhaust pressures, engine speed, compression ratio, valve timings and exhaust system dynamics [79].

To show the influence of the residual gas fraction on particle number emissions, a variation of the valve overlap was performed by varying the exhaust open timing. The exhaust back-pressure was set to two different values: in the first place equal to the charge air pressure to simulate the back-pressure increase of a turbocharger. The second setting was dethrottled and thus simulating boosting by a supercharger. For the variation, the injection timing of the high flow injector was set to PN optimised timing at -290 °CA $_{\text{aTDCf}}$ with an injection pressure of 20 MPa and kept constant. The charge air pressure was adjusted to keep the IMEP constant at 1.4 MPa and the exhaust gas air-fuel ratio to $\lambda = 1$. The ignition timing was set to knock-limited spark advance. Extracts of these investigations were published at the Conference on Diesel- and Gasoline Direct Injection, 2014, Berlin [19].

For these investigations, the charge motion inlays to generate a tumble (60 %) and swirl were compared to the baseline case. In Figure 5.12, the emitted PN concentrations of the variations are plotted versus the exhaust open timing. As the intake open timings were fixed, the valve overlap increases for later exhaust open timings. Noticeable is the higher PN emissions using the baseline configuration. This is in accordance with the findings of section 5.1. For the baseline configuration, the increase of the exhaust back-pressure[5], and thus increased residual gas fraction, leads to reduced particle emissions. For the tumble and the swirl case, the reduction of the PN concentration by the increased exhaust back-pressure, and thus higher residual gas fraction, is on a lower level. Using the earliest exhaust open timings with close to zero valve overlap, the PN concentration is even slightly increased.

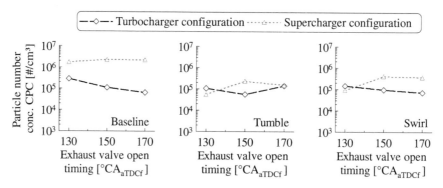

Figure 5.12: PN concentration for the variation of the exhaust valve timings and different exhaust gas back-pressures at 2000 rpm, 1.4 MPa IMEP, SOI: -290 °CA, $p_{Inj.}$: 20 MPa, high flow injector, ignition timing: knock-limited spark advance

The combustion data is shown in Figure 5.13. The CoV of IMEP is on a higher level for most of the cases with increased exhaust gas backpressure. This can be attributed to the later centre of combustion (MFB50%). As the turbulence decreases after TDC, the fluctuations of the inflammation and the combustion increase with a later MFB50%, thus increasing the cycle-to-cycle variations. The flame development angle is more strongly influenced by the charge motion strategy than by the residual gas fraction. With increased in-cylinder charge motion, the flame development angle can be shortened by a better mixture homogenisation and increased turbulence. Also noticeable is that the MFB50% needs to be shifted late for the large valve overlap, whereas the ignition timing can be advanced with the reduced exhaust back-pressure and large valve overlap.

Operating with a large valve overlap and a positive scavenging pressure ratio[6], the residual gas fraction is on a low level because of a good scavenging process, as shown in Figure 5.14. With increased exhaust gas back-pressure, the scavenging process

---

[5]supercharger configuration: dethrottled; turbocharger configuration: equal to charge air pressure
[6]charge air pressure > exhaust back-pressure

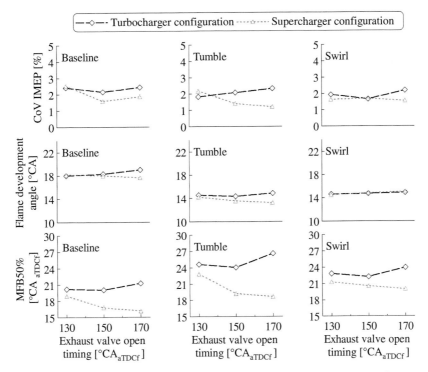

Figure 5.13: Combustion characteristics for the variation of the exhaust valve timings and different exhaust gas back-pressures at 2000 rpm, 1.4 MPa IMEP, SOI: - 290 °CA, $p_{Inj.}$: 20 MPa, high flow injector, ignition timing: knock-limited spark advance

gets restricted and the residual gas fraction increases. Therefore, the residual gas concentration is higher for operating with an increased exhaust gas back-pressure. A possible explanation for the reduced PN emissions with larger valve overlap (later exhaust valve open timing) for the turbocharger configuration is the increased residual gas concentration caused by the reduced scavenging efficiency. For the dethrottled operation however, the residual gas concentration is reduced with increased valve overlap.

The exhaust gas temperature can be influenced by the exhaust back-pressure, by the timing of the centre of combustion as well as the exhaust valve open timing (evo). With later evo, the loss of expansion work is reduced and thus the temperature of the out flowing exhaust gas is on a lower level. By raising the exhaust gas back-pressure, the temperature in the exhaust runners is also increased. An advance of 20 °C to 40 °C was measured for the three charge motion strategies. Because the engine was more prone to knocking at operation with the tumble inlay and the turbocharger configuration, the MFB50% needed to be shifted late. Therefore, the exhaust gas temperature is the highest for this combination. As the baseline configuration showed the most knocking resistance, the MFB50% is the earliest and thus the exhaust gas temperatures for both exhaust gas back-pressure configurations was on the lowest level.

Based on the higher exhaust gas temperatures, the potential for post-oxidation is higher for the turbocharger configuration. Therefore, the emission of hydrocarbons is on a lower level for the turbocharger configuration. The THC emission is thereby nearly independent of the charge motion strategy. For the turbocharger configuration, the THC emissions are nearly uninfluenced by the exhaust valve open timing. The supercharger configuration shows a clear tendency of increased THC emissions for later exhaust valve open timings. These can be attributed to the scavenging effect[7] and thus slightly rich air-fuel ratio inside the combustion chamber. As shown in section 5.2, a slight enrichment does not significantly increase the PN emissions, whereas the hydrocarbon emissions are strongly influenced.

In Figure 5.15 the particle size distributions for the baseline configuration and the exhaust valve open timings (evo) 130 °CA $_{aTDCf}$ and 170 °CA $_{aTDCf}$ are shown. For the two exhaust valve settings with supercharger configuration, both the accumulation and the nucleation mode particles are increased. As both the accumulation mode and the nucleation mode particles are increased, it is assumed that the higher PN emissions result from increased soot emissions. Therefore, differences in the mixture formation process are likely.

To get further information about the mixture preparation, the GT Power TPA-Model was used. In Figure 5.16 the calculated in-cylinder temperature of the four measurement points shown in Figure 5.15 are plotted. Additionally the exhaust valve lifts are plotted on the secondary vertical axis. The temperature drops at earlier degree crank angle for evo 130 because of the opening of the exhaust valves. After the exhaust valve of the evo 170 operation point opened, the temperature inside the cylinder drops to the same level compared to the evo 130 cases. The temperature at start of injection and during the injection process are important for the evaporation process and the mixture formation. It is noticeable, that the temperatures for both supercharger cases at SOI timing (430 °CA $_{aTDCf}$, zoomed in the data loupe) are on a lower level compared to the turbocharger cases. Also the order of the temperatures is inversely proportional to the PN concentrations measured. This indicates that the higher temperatures and thus increased evaporation potential support the mixture formation process and reduce the PN concentrations.

Summed up, these findings fit well to the investigations by Xu et al., who investigated the influence of valve timings and residual gas fraction on the combustion characteristics of a PFI engine [207]. They used PIV measurements as well as chemiluminescence measurements and a heat release analysis to show the influence on the flow pattern,

---

[7]During valve overlap both intake and exhaust valves are open. With higher pressure in the intake compared to the exhaust port, charge air can flow directly from the intake to the exhaust port without taking part in the combustion process. Thus the oxygen concentration is increased in the exhaust, whereas the in-cylinder air-fuel ratio is slightly rich for exhaust stoichiometric.

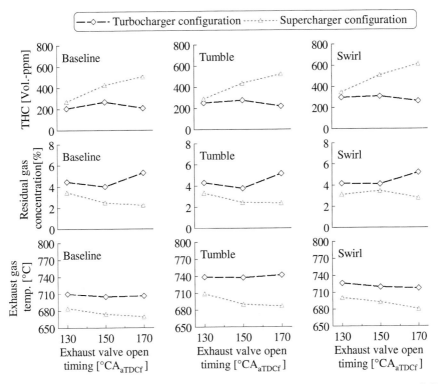

Figure 5.14: Exhaust gas emissions for the variation of the exhaust valve timings and different exhaust gas back-pressures at 2000 rpm, 1.4 MPa IMEP, SOI: -290 °CA, $p_{Inj.}$: 20 MPa, high flow injector, ignition timing: knock-limited spark advance

flame propagation and combustion. Xu et al. state that the change of exhaust valve timings mainly affected the smaller vortex structure rather than the larger one. The slightly advanced exhaust valve open timings resulted in a shorter ignition delay and combustion duration due to the thermal state caused by the higher trapped residual gas fraction. However, further advancing the exhaust valve timings leaded to retarded combustion timings and higher cyclic variations due to the higher dilution level and lower turbulent kinetic energy.

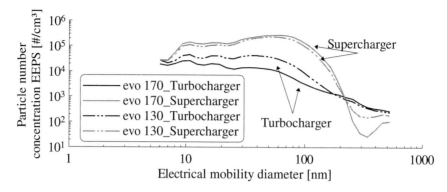

Figure 5.15: Particle size distribution for the variation of the exhaust valve timings and different exhaust gas back-pressures with baseline configuration at 2000 rpm, 1.4 MPa IMEP, SOI: -290 °CA, $p_{Inj.}$: 20 MPa, high flow injector, ignition timing: knock-limited spark advance

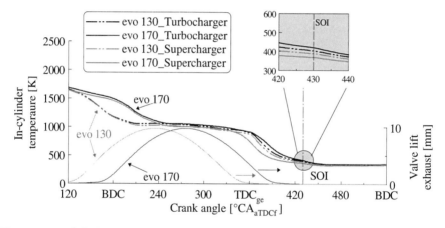

Figure 5.16: Calculated temperatures inside the cylinder for the variation of the exhaust valve timings and different exhaust gas back-pressures with baseline configuration at 2000 rpm, 1.4 MPa IMEP, SOI: -290 °CA, $p_{Inj.}$: 20 MPa, high flow injector, ignition timing: knock-limited spark advance

5.4 Effect of external, high pressure EGR on PN emissions

The discussion of previous results on the influence of EGR on PN emissions in gasoline engines showed no clear tendency, as discussed in section 2.4.5. Therefore, it was decided to get a closer look to the mechanisms of EGR. The engine was equipped with a high pressure EGR system and an inter-cooler. The secondary circuit of the inter-cooler could be operated with engine cooling at 90 °C or with cooling water at 20 °C respectively. Operation with engine cooling is thereby representing actual setups in automotive applications, whereas the operation with cool water is used to show the potential of cooled EGR. To increase the EGR fraction, the exhaust gas backpressure needed to be increased due to the high-pressure EGR setup. In the first part of this section, the comparison of the hot and cold EGR cooling is shown. The second part of the section describes the combined influence of EGR and charge motion on engine performance and emissions.

5.4.1 Influence of EGR cooling

The charge air temperature was kept constant at 30 °C for the previously shown results. As the EGR inter-cooler was operated with engine cooling at 90 °C, the EGR was hotter compared to the charge air. Therefore, the temperature of the charge in the intake channel was increased with higher EGR rates. Whereas operating the inter-cooler with cold water did not increase the charge air temperature. To show the influence of the EGR temperature, an EGR ratio variation was performed at constant engine operation at 2000 rpm and 1.4 MPa IMEP. Thereby the MFB50% was adjusted to knock-limited spark advance at 0 % EGR and then kept constant for the increased EGR rates. The tumble inlay with 60 % closed intake channel was used. The valve timings were set to early exhaust open timing and late intake open timing, resulting in zero valve overlap. The high flow injector was operated with a constant injection timing of -290 °CA $_{aTDCf}$ and 20 MPa injection pressure. The resulting PN emission for these variations is shown in Figure 5.17. It is obvious that using the hot engine cooling, the EGR tolerance of the engine is lower and thus it is only possible to operate with an EGR fraction of up to about 11 % due to rough engine operation (CoV IMEP > 4 %) and invreased charge air temperature (above 46 °C). For the cold water, it is possible to use up to 20 %. Thereby the PN emissions are on a low level without using EGR, and the PN emissions are not significantly increased in both cases.

Taking a look at the engine data in Figure 5.18, a steep increase of the charge air temperature for the hot EGR case is obvious. By mixing 10 % of the hot EGR to the charge air, the temperature in the intake runners rises up to about 46 °C. Using the cold cooling water in the EGR inter-cooler, the charge air temperature can be kept constant at 30 °C. The hydrocarbon emissions are slightly increased by higher EGR fractions.

The combustion stability (CoV IMEP) is negatively affected by higher EGR rates, both for hot and cold EGR. Due to the diluting effect of the EGR, the flame development angle as well as the rapid burn angle are prolonged inversely with the EGR mass fraction. The effect of the hot EGR on both flame development angle is stronger compared to the cold EGR.

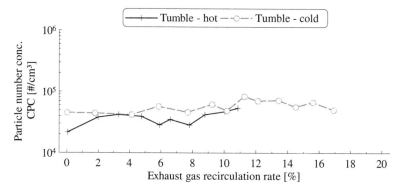

Figure 5.17: PN emissions for the EGR variation with cold and hot cooling at 2000 rpm, 1.4 MPa IMEP, SOI: -290 °CA, $p_{Inj.}$: 20 MPa, high flow injector, ignition timing: set for MFB50% at 27 °CA $_{aTDCf}$

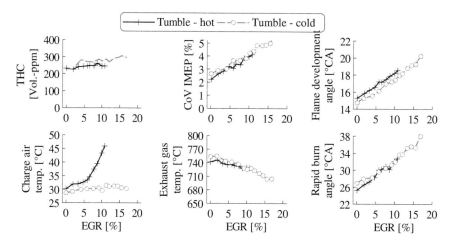

Figure 5.18: Combustion characteristics for the EGR variation with cold and hot cooling at 2000 rpm, 1.4 MPa IMEP, SOI: -290 °CA, $p_{Inj.}$: 20 MPa, high flow injector, ignition timing: set for MFB50% at 27 °CA $_{aTDCf}$

Additional data sets of the combustion process calculated with the GTPower TPA model (described in section 3.3) are shown in Figure 5.19. In the upper graphs, the calculated in-cylinder temperature at injection, the heat release rate (ROHR) and the temperature of the burned zone during combustion are plotted for four EGR rates of the hot EGR case. The same data is shown in the lower graphs for the cold EGR case. It is obvious, that the in-cylinder temperature during injection is increased for the hot EGR case with higher EGR fractions. Higher temperatures in the cylinder during the injection process increase the evaporation potential and thus enhance the mixture formation process. On the other hand, higher temperatures during injection lead to higher temperatures during compression and thus promote engine knocking. As the temperature is increased, the ignition delay is not as strongly influenced by the hot EGR compared to the cold EGR case. For the cold EGR case, the ignition delay is strongly prolonged and thus the ignition timing needs to be advanced to keep MFB50% constant. Furthermore, the peak temperatures for both hot and cold EGR are reduced using higher EGR rates. Thus the temperature inside the cylinder drops earlier to lower levels thus reducing the post oxidation potential. Furthermore, due to the addition of inert gas the local oxygen concentration is reduced.

Summed up, EGR has positive and negative effects on PN formation and oxidation. For the experiments shown in this section, these effects are just about balanced. The PN concentration is just slightly increased for higher EGR rates. Nevertheless EGR has additional benefits concerning exhaust gas emissions and fuel consumption. These benefits can be used without increasing the PN emissions.

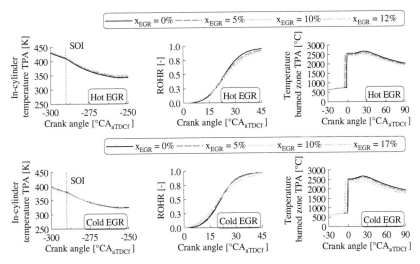

Figure 5.19: Additional thermodynamic analysis for the EGR variation with cold and hot cooling at 2000 rpm, 1.4 MPa IMEP, SOI: -290 °CA, $p_{Inj.}$: 20 MPa, high flow injector, ignition timing: set for MFB50% at 27 °CA $_{aTDCf}$

5.4.2 EGR and charge motion

As shown in the previous section, EGR effects the mixture formation and the combustion process. For further investigations of the effect of cooled EGR, measurements with different charge motion strategies and thus different turbulence level in the cylinder were performed. Once again the engine was operated at boosted operation at 1.4 MPa

IMEP and an engine speed of 2000 rpm. The valve timings were set to early exhaust open timing and late intake open timing, resulting in zero valve overlap. The injection pressure was set to 20 MPa and the SOI of the high flow injector to -290 °CA $_{\text{aTDCf}}$. The spark advance was set to a constant MFB50% of 22 °CA $_{\text{aTDCf}}$. The boost pressure was adjusted to a constant IMEP at an air-fuel ratio of 1.0.

The measured PN concentrations using the baseline, the 60 % tumble and the swirl configuration are shown in Figure 5.20. The different EGR tolerance for the three charge motion strategies is obvious. Just like in the previous section, a clear trend for PN emissions was not found. Using the baseline configuration, the PN emissions even without EGR are on a higher level. This was discussed in section 5.1.1. However, for increased EGR fractions up to 20 %, the PN emissions are increased. As discussed in the previous section, using the tumble inlay, the PN emissions are not significantly influenced by using EGR. With lower EGR fractions, the PN emissions are slightly increased. Operating with higher EGR rates, the PN emissions can be reduced by applying EGR. The swirl configuration shows the highest EGR tolerance. EGR fractions of up to about 30 % are possible. With EGR fractions of up to about 15 %, the PN emissions are slightly increased. Further increasing the EGR fraction leads to reduced PN concentrations.

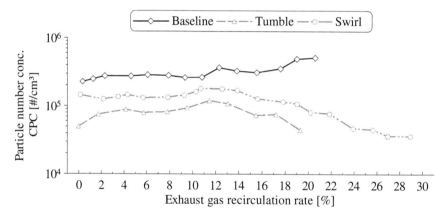

Figure 5.20: PN emissions for the EGR variation using the tumble and the swirl configuration at 2000 rpm, 1.4 MPa IMEP, SOI: -290 °CA, $p_{\text{Inj.}}$: 20 MPa, high flow injector, ignition timing: set for MFB50% at 22 °CA $_{\text{aTDCf}}$

Additional exhaust gas analysis and combustion data are shown in Figure 5.21. As shown in section 5.1.1, the nitrogen oxides are as well influenced by the charge motion strategy as the flame development angle, the rapid burn angle and thus the exhaust gas temperature. The injected fuel mass depends on the charge motion strategy and the running smoothness of the engine (CoV IMEP).

Operating without EGR, the baseline case emits the lowest amount of hydrocarbons. Applying EGR in general dilutes the charge, reduces the flame speed and thus slows down the burn rate. In some cycles, combustion is not completed by the time the exhaust valves open. With increased EGR fraction, the hydrocarbon emissions are strongly increased for the baseline and the swirl case, whereas the hydrocarbon emissions for the tumble case are only slightly increased. To increase the EGR fraction, a higher exhaust gas pressure is needed. Even though the exhaust gas pressure is increased, the exhaust gas temperature can be reduced using EGR. This is caused by the reduced in-cylinder temperatures using EGR. A reduction of the engine running smoothness, caused by slow burning cycles and even misfires, goes along with the dilution by EGR. As shown in section 5.1.1, the charge motion inlays have various degrees of impact on the mixture formation. Both the swirl and the tumble motion increase the flame development angle by the increased turbulence at injection timing and compression. However, during the combustion process, the tumble motion dissipates and does not increase the burn rate and thus the rapid burn angle as much as the swirl motion. Applying EGR to the system both increases the flame development angle and the rapid burn angle.

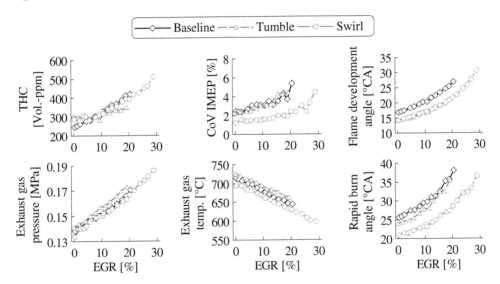

Figure 5.21: Combustion characteristics for the EGR variation using the tumble and the swirl configuration at 2000 rpm, 1.4 MPa IMEP, SOI: -290 °CA, $p_{Inj.}$: 20 MPa, high flow injector, ignition timing: set for MFB50% at 22 °CA $_{aTDCf}$

For these variations, additional analysis using the TPA model (described in section 3.3) was performed. In Figure 5.22, the results of the EGR variations for the three charge motion strategies are plotted. The baseline case is plotted in the upper row, the tumble case in the middle row and the swirl case in the lower row. It is obvious that the temperature before injection depends on the EGR ratio. For higher EGR rates the peak temperature in the cylinder and the temperature at the end of combustion are lower, thus reducing the wall heat losses. In combination with the dilution effect of EGR, the burn rate is significantly reduced with higher EGR rates. Therefore, the spark timing needs to be significantly advanced for a constant MFB50%. Furthermore, the burning duration is prolonged and thus the end of combustion is shifted towards 'late'. As already shown in the previous section 5.4.1, the peak temperatures as well as the temperature at the end of combustion are reduced. Therefore, the potential for a post oxidation of the soot is lower using EGR.

Summed up, EGR shows benefits concerning the in-cylinder peak temperatures and thus the formation of nitrogen oxides as well as on engine knocking. However, caused by the lower temperatures, the evaporation potential is lower and in combination with the dilution by EGR, the mixture formation process is more challenging. Also the potential for post oxidation of soot is lower because of the lower exhaust gas temperatures. For the engine used in these investigations, the effects influencing PN emissions are more or less evenly balanced. The PN emissions are not significantly increased using EGR, while the benefits of EGR, such as lower fuel consumption and less NO_x emissions, can be used. As the EGR slows down the burn rate, applying a large scale charge motion can compensate the lower burn rate and enhance the mixture formation process as well as the EGR tolerance of the combustion process.

Additional high-speed measurements were performed using the tumble and the swirl configuration, shown in Figure 5.23. The cases without EGR are identical to the ones shown in section 5.2 for the $\lambda = 1$ case. The graphs on the right hand side show measurements with 10 % EGR. Once again the standard deviation of the intensity is plotted versus the crank angle for the 80 cycles measured. The reduced maximum values for both EGR cases are significant. This is assumed to be caused by the reduced temperatures during the combustion process, because the intensity of gas and soot luminosity are a function of the temperature. Also noticeable are the lower variations for late degree crank angle for the tumble configuration with and without EGR. For the swirl case, single cycles with higher intensities at late degree crank angles were measured. These are indicators for diffusive combustion, causing high soot emissions. As for the tumble configuration, the PN emissions are on a lower level compared to the other charge motion strategies, as shown in Figure 5.20.

Images of the interesting cycles of both measurements using the swirl configuration are shown in Figure 5.24. The late burning cycle (dark grey, dashed) was already discussed in section 5.2. As already discussed for the air-fuel ratio variation, the diffusive combustion for the operation without EGR using the swirl configuration is assumed to be bound to the exhaust valve gap. This can be caused for instance by a fuel-wall interaction during the mixture formation process.

Taking a look at the EGR case, a bright diffusive flame is visible on the left hand side. As the diffusive combustion is not directly bound to the combustion chamber walls and moves downwards with the piston, it is assumed that the diffusive combustion was caused by an inhomogeneous mixture in the gas phase.

As the same phenomena was visible for the air-fuel variation, and is not visible for the tumble configuration, it is assumed, that these inhomogeneities are caused by the

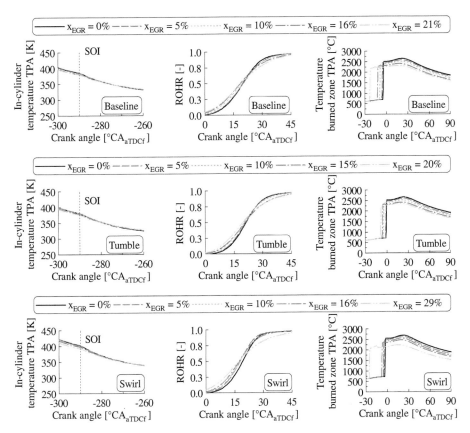

Figure 5.22: Additional thermodynamic analysis for the EGR variation using the tumble and the swirl configuration at 2000 rpm, 1.4 MPa IMEP, SOI: -290 °CA, $p_{Inj.}$: 20 MPa, high flow injector, ignition timing: set for MFB50% at 22 °CA $_{aTDCf}$

generation of the charge motion or an incomplete mixture formation process. As the swirl charge motion was generated by closing one intake channel, without deactivating the intake valve, it is assumed that an interaction of the spray and the intake valve can cause mixture inhomogeneities. Because no valve was deactivated, the intake valve was open during the injection process, but no charge air was flowing through this intake channel. Therefore, it is assumed that the impinged fuel on the intake valve does not interact strongly with the in-cylinder charge motion, and thus tends to build fuel-rich zones, which are one cause for diffusive combustion.

To prove this assumption, additional measurements using a CCD camera (Imager Compact by LaVision) were performed during the injection process. A strong spray-valve interaction was visible in some cycles, as exemplary shown for the baseline and the swirl case in Figure 5.25. The maximum frame rate of the CCD camera limits

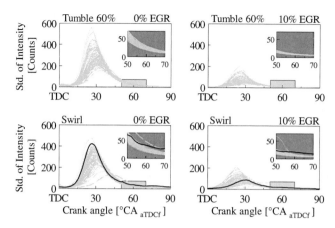

Figure 5.23: Analysis of high-speed images for the EGR variation using the tumble and the swirl configuration at 2000 rpm, 1.4 MPa IMEP, SOI: -290 °CA, $p_{Inj.}$: 20 MPa, high flow injector, ignition timing: set for MFB50% at 22 °CA $_{aTDCf}$

the number of pictures taken to one single image per cycle. However, this picture is taken in a higher resolution compared to the images taken with the high-speed camera. Therefore, it is possible to show small-scaled effects like the spray-valve interaction. 200 cycles were measured with each charge motion strategy. The camera timing was set to 15 °CA after SOI, so that the spray was formed properly. The open intake valve is visible on the left hand side of the images. These three figures chosen are representative for the baseline, the tumble and the swirl case. For the tumble case, a spray valve interaction with splashing droplets like for the baseline and the swirl case was not observed. For the swirl configuration, a strong spray-valve interaction comparable to the baseline case was noticeable in several cycles.

To show this effect in terms of statistics, the following post-processing was implemented: The area of the intake valve, where the interaction takes place was cut out of the image, shown in the grey boxes in Figure 5.25. For cycles without spray-valve interaction, no fuel droplets are observed in this area. A threshold value of 500 counts was used to reduce the background noise. The intensities of all the pixels in this grey marked area was used to calculate the standard deviation of intensity. Comparable to the previously shown high-speed measurements, the standard deviation of intensity is sensitive to small luminous spots. As the liquid fuel scatters light, according to the Mie-theory discussed in section 3.4.1, splashing droplets show high intensities and the cycles with spray-valve interaction can thus be distinguished from cycles without spray-valve interaction. The results of this calculation are shown in Figure 5.26.

The standard deviation of intensity for the calculation window of the baseline case is plotted versus the cycle number in the upper graph, the tumble case is shown in the middle graph and the swirl case in the lower graph. The lowest average value and the low number of peaks with higher value is significant for the tumble case. Additionally

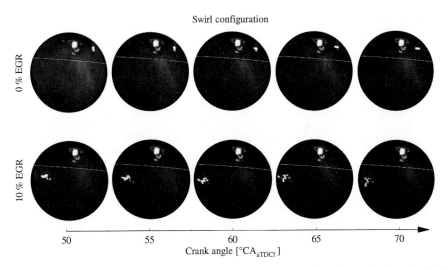

Figure 5.24: Highlighted Cycles of Figure 5.23 using the swirl-configuraion for the EGR variation at 2000 rpm, 1.4 MPa IMEP, SOI: -290 °CA, $p_{Inj.}$: 20 MPa, high flow injector, ignition timing: set for MFB50% at 22 °CA $_{aTDCf}$

the maximum value of the highest peak is comparatively low. For the baseline and the swirl case on the other hand, the average value is higher. Furthermore, a higher number of cycles shows high values, thus indicating cycles with a strong spray-valve interaction with splashing droplets.

These results indicate that the higher PN emissions caused by the baseline configuration and in some cases by the swirl configuration, can be attributed to a strong spray-valve interaction. It is assumed that by using the tumble inlay, the flow velocities on the upper part of the intake valves are increased because of the reduced flow area. Higher flow velocities support the spray break-up and therefore support the mixture formation process. Additional endoscopic measurements using particle image velocimetry are planned for the follow-up project to measure the flow velocities inside the combustion chamber and thus prove this assumption.

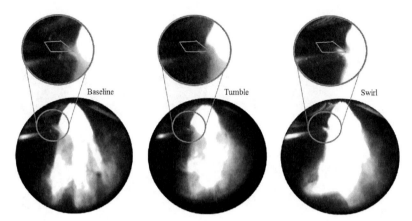

Figure 5.25: Images taken with a CCD camera 15 °CA after SOI for the baseline (left), tumble (middle) and swirl case (right) at 2000 rpm, 1.4 MPa IMEP, SOI:-290 °CA, $p_{Inj.}$: 20 MPa, high flow injector, ignition timing: set for MFB50% at 22 °CA $_{aTDCf}$

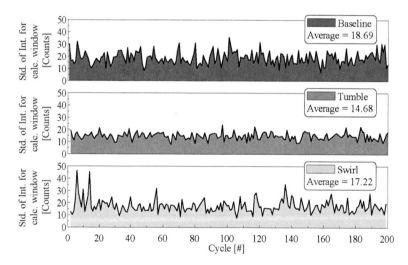

Figure 5.26: Statistical analysis of the spray-valve interaction at 2000 rpm, 1.4 MPa IMEP, SOI:-290 °CA, $p_{Inj.}$: 20 MPa, high flow injector, ignition timing: set for MFB50% at 22 °CA $_{aTDCf}$

5.5 Influence of injector static flow and injection pressure

Several SOI variations at 2000 rpm were done to show the influence of the reduced static flow. The engine load was chosen based on relevant points of the NEDC test cycle. An engine load of 0.2 MPa IMEP is equal to a constant driving at 50 km/h. An operation with nearly WOT and an IMEP of 0.8 MPa is representative for an acceleration up to 120 km/h. For these first tests, the ignition timing was adjusted for a thermodynamically optimised MFB50 % at 8 °CA aTDCf. The third operation point chosen was at boosted operation. Large cars are equipped with smaller engines to operate at specific higher power and thus at engine operation with a higher specific efficiency. The reduction in displacement goes along with a reduction in mass and friction losses as well as a reduction in throttling losses due to operating at higher specific load for a larger part of the operating range. This approach is known as 'Downsizing'. To compensate the reduced displacement, these engines use either increased compression ratios or boost pressures to improve the volumetric efficiency [62].

The two injectors used for this comparison were combined with the three large scale charge motion configurations (baseline, tumble (60 %) and swirl). The experimental matrix is visualised in Figure 5.27. The specific data of the injectors are shown in section 3.2.2.

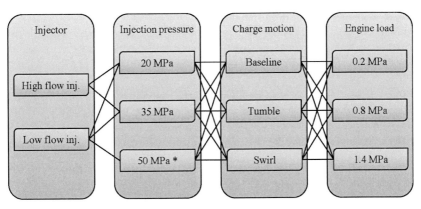

* Exclusively with low flow injector

Figure 5.27: Experimental matrix for the investigations on the influence of the fuel pressure and the injector hydraulic flow on PN emissions

In order to show the influence of the reduced static flow, a comparison of the two injectors with equal injection pressure of 35 MPa is shown in the following sections. The injection timing for the low flow injector needed to be prolonged to inject the same amount of fuel. The influence of an increased injection pressure is shown later in section 5.5.4.

5.5.1 Influence of injector static flow at low engine load

The first operation discussed is the throttled operation at 2000 rpm and an IMEP of 0.2 MPa, which equals constant driving at 50 km/h in the NEDC. The valve timings were chosen according to a reference engine. Thereby the maximum intake valve lift was reduced to 5 mm. The intake pressure was reduced with the throttle to reduce the intake mass flow. The air-fuel ratio was adjusted to $\lambda = 1$ and the ignition timing was set to an MFB50% of 8 °CA $_{aTDCf}$. As the low flow injector has a lower hydraulic flow than the high flow injector, the injection duration needs to be prolonged to inject the same amount of fuel with 35 MPa injection pressure.

The PN concentrations emitted by the three charge motion configurations using the two injectors is shown in Figure 5.28. The baseline configuration is shown in the first column, while the tumble configuration is shown in the second column and the swirl configuration in the third column. Noticeable is that the PN emissions for all three charge motion strategies are on a low level for most part of the variation. Only for the early injection timing, an increase of particle number emissions is measurable. This increase is assumed to be driven by an interaction of the liquid fuel and the piston. For these early injection timings, the free distance between injector tip and piston surface is low and thus an impingement of liquid fuel on the piston is possible [174]. The liquid fuel on the piston surface is not evaporated fast enough and burns in a diffusive flame, thus emitting soot [193].

Slightly different PN concentrations were measured using the low flow injector in combination with the tumble and the baseline configuration respectively. As the PN emission level is on a low level (below 10^5 #/cm³), this difference is assumed to be an artefact of the measurement devices. Caused by the low exhaust gas mass flow, the specific particle emission at this operation point is low and the influence on the overall PN emissions in a test cycle is low.

The particle size distributions of SOI = -280 °CA $_{aTDCf}$ (highlighted as 1) are plotted for the three charge motion strategies and the two injectors in Figure 5.29. It is obvious that independent of the charge motion strategy and the injector hydraulic flow, the number of particles with a diameter smaller than 23 nm is on a high level. The cause of these small nucleation mode particles is not yet known. It is assumed that nucleation mode particles consist of primary soot particles, metal oxide particles of fuel or oil additives as well as volatile compounds [60, 61, 131]. For GDI engines, the size of primary soot particles was found to be in the range of 10 nm to 30 nm [52, 152], which is quite comparable to Diesel engines. However, Giechaskiel states that the distribution can be wider for GDI engines [60]. Thus a larger amount of particles with a diameter smaller than 23 nm can exist. Compared to Diesel engines, the structure of the particles can be more amorphous and more unburned hydrocarbons or volatile

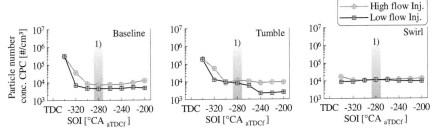

Figure 5.28: PN emissions for the SOI variations with high flow and low flow injector at 2000 rpm, 0.2 MPa IMEP, $p_{Inj.}$: 35 MPa, ignition timing: set for MFB50% at 8 °CA $_{aTDCf}$

organics can be found. This means that differences in the thermal pre-treatment[8] might influence the measurement result.

Maier et al. compared the emission of particles smaller than 60 nm from gasoline and hydrogen operation from a single cylinder GDI engine [121]. They found comparable number concentrations of particles with a diameter below 10 nm for low engine load of 0.2 MPa IMEP at 2000 rpm, which they assume to be due to the identical formation process. The inlet air could not account for these number concentrations and the volatile particles were removed by the VPR. Thus they state that abrasion from the cylinder liner or metal additives from the lubricant remain as possible origins for the sub 23 nm particles.

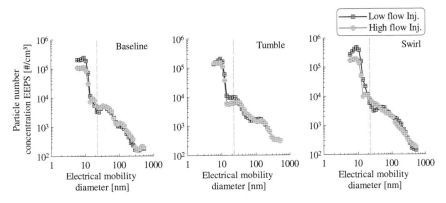

Figure 5.29: Particle size distribution for the SOI variations with high flow and low flow injector at 2000 rpm, 0.2 MPa IMEP, $p_{Inj.}$: 35 MPa, ignition timing: set for MFB50% at 8 °CA $_{aTDCf}$

---

[8]temperature and residence time of the exhaust gas in the PN system

As shown in Figure 5.30, the running smoothness of the engine strongly depends on the injection timing. With the later start of injection, the cycle-to-cycle changes rise (shown as CoV IMEP). The charge motion strategies affect neither the running smoothness nor the emission of hydrocarbons nor the exhaust gas temperature. This indicates that for these low charge air mass flows, no stable charge motion is formed inside the cylinder using the inlays. The hydraulic flow of the injector does not influence the combustion process strongly at this low engine load point.

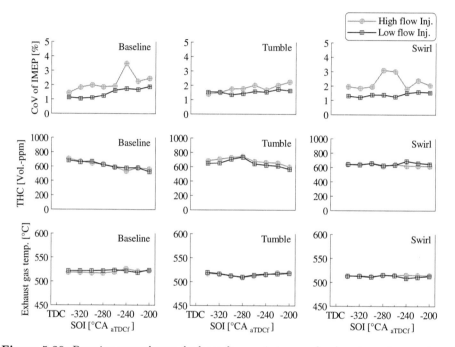

Figure 5.30: Running smoothness, hydrocarbon emissions and exhaust gas temperatures for the SOI variations with high flow and low flow injector at 2000 rpm, 0.2 MPa IMEP, $p_{Inj.}$: 35 MPa, ignition timing: set for MFB50% at 8 °CA $_{aTDCf}$

Summed up, the difference in hydraulic flow did not lead to measurable differences in the PMP-conform particle measurement at low engine load. However, concerning sub-23 nm particles, a slight increase was measurable with reduced injector hydraulic flow. As the exhaust gas flow is low for this low engine load and speed operation, the slight increase in sub-23 nm particles is not assumed to increase the overall PN emissions in a test cycle significantly.

5.5.2 Influence of injector static flow at WOT operation

The second operating point to compare the different hydraulic flow is at WOT operation with an indicated mean effective pressure of 0.8 MPa. For these tests, the ignition
timing was adjusted for a thermodynamically optimised MFB50 % at 8 °CA $_{aTDCf}$. The
air-fuel ratio was kept constant at stoichiometry ($\lambda = 1$). The injection pressure used
with both injectors was 35 MPa. Therefore, the injection timing for the low flow injector needed to be prolonged to inject the same amount of fuel. The valve timings
were chosen according to a reference engine. Thereby the maximum intake valve lift
was reduced to 5 mm and the valve timings set to a slight negative valve overlap.
Therefore, no effects of scavenging should influence the results.

Emission characteristics

The PN concentrations measured for these six SOI variations are shown in Figure 5.31.
In the graph on the left hand side, the engine was operated with the baseline configuration without additional large scale charge motion. The tumble configuration is
shown in the middle graph and the swirl configuration on the right hand side. For
both injectors a characteristic PN-emission behaviour for GDI engines is obvious with
lowest PN concentrations at SOI between -300 °CA$_{aTDCf}$ and -260 °CA$_{aTDCf}$.
Injecting the fuel earlier leads to an interaction of the spray and the piston, caused
by the short distance between the piston close to TDC and the injector tip. Thus
the probability of a liquid fuel film on the piston increases with earlier injection timings. The liquid fuel can not be prepared sufficiently during the mixture formation
process, resulting in a diffusive combustion on the piston surface causing high soot
emissions [41]. This phenomenon is known as 'Poolfire' [193].
On the other hand, injecting the fuel later increases the particle number concentration
because of the reduced time for the mixture formation process. High cyclic variations
are a consequence of this short time for mixture preparation, shown in the upper row
of Figure 5.32.
Even though a higher penetration length was measured in the pressurised chamber
using the low flow injector, as shown in section 4.3, the PN emissions are not increased
compared to the high flow injector. At optimised injection timings, the PN emissions
using the baseline configuration are on a similar level for the two static flows. However,
for earlier and later injection timings, a difference in PN concentrations is obvious for
the two injectors with higher PN concentrations using the high flow injector. Because
of the higher flow, the spray of the high flow injector penetrates faster. Therefore, a
stronger interaction of spray and piston is assumed for the earlier injection timings,
associated with increased piston wetting and thus diffusive combustion. Measurements
by Disch et al. [40] showed that under transient engine operation even single cycles
with early injection timings and thus fuel impingement on the piston, increase the
overall PN emissions significantly. These findings apply well to the findings of Stumpf
et al. [183].

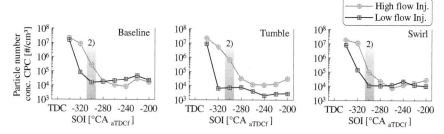

Figure 5.31: PN emissions for the SOI variations with high flow and low flow injector at 2000 rpm, 0.8 MPa IMEP (WOT), $p_{Inj.}$: 35 MPa, ignition timing: set for MFB50% at 8 °CA$_{aTDCf}$

The reduced static flow of the low flow injector leads to smaller droplets and thus an improved surface-to-volume ratio. This leads to a faster evaporation and better mixture formation and can be assumed to be the reason for the lower PN concentrations for the late injection timings.

By inducing a tumble motion in the cylinder, the PN concentrations using the low flow injector can be significantly reduced for all SOI. For the high flow injector, the PN emissions are increased for all injection timings. However, for the implementation of a swirl motion in the cylinder, the results are different compared to the tumble configuration. The high flow injector shows lowest minimum PN concentrations, while the PN concentration using the low flow injector is slightly reduced compared to the baseline case.

The running smoothness of the engine (CoV of IMEP) and the hydrocarbon emissions (THC) of the discussed SOI variations are plotted in Figure 5.32. The running smoothness is neither influenced significantly by the variation of the static flow of the injector, nor by the implementation of a large scale charge motion. It depends on the injection timing and thus the time for mixture preparation. The THC emission is also independent of injector characteristics and large scale charge motion at WOT, except for the early injection timings. Using the high flow injector, the THC emission is significantly increased compared to the low flow injector. This goes along with the significantly higher PN emission and can be associated to a stronger piston wetting.

Analysis at injection timing -300 °CA$_{aTDCf}$

For further analysis of the shown results, additional data at SOI = -300 °CA$_{aTDCf}$ (highlighted as 2) in Figure 5.31) are shown in the following section. The particle size distribution is shown in Figure 5.33. The size distribution was measured using the EEPS. The exhaust gas is prepared by the VPR of the CPC. Thus volatile particles should be removed. For all three charge motion strategies the PN emissions are sig-

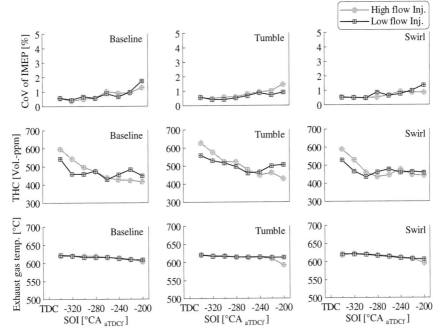

Figure 5.32: CoV of IMEP and hydrocarbon emissions (THC) for the SOI variations with high flow and low flow injector at 2000 rpm, 0.8 MPa IMEP (WOT), $p_{Inj.}$: 35 MPa, ignition timing: set for MFB50% at 8 °CA $_{aTDCf}$

nificantly increased using the high flow injector. This increased PN concentration is caused by a higher amount of accumulation particles. Increased accumulation mode particles are associated with a diffusive combustion and soot formation [102], while nucleation mode particles are associated with primary soot particles, ash and metallic wear [60, 61, 131]. Therefore, problems with mixture preparation and following fuel rich zones are assumed using the high flow injector and the tumble configuration.

To explain these results, high-speed imaging measurements were taken at SOI = -300 °CA$_{aTDCf}$ for both injectors and all charge motion configurations. To visualise the results of the 300 cycles, the average intensity and the standard deviation of the intensity (std. of intensity) for each degree crank angle were calculated. In Figure 5.34, the std. of intensity for the 300 cycles is plotted versus the crank angle for both injectors (top: low flow; bottom: high flow) and the three charge motion strategies. The lower maximum values for the swirl configuration and the high flow injector were caused by deposits on the optical access. The highest intensities and highest std. of intensity are measured during the main combustion close to the top dead centre (TDC). For the low flow injector, the variance of the 300 cycles is lower compared to the high flow injector in all cases. Especially after the end of the regular combustion

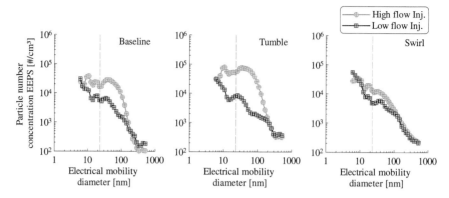

Figure 5.33: Particle size distribution at -300 °CA$_{\text{aTDCf}}$ at 2000 rpm, 0.8 MPa IMEP (WOT), p$_{\text{Inj.}}$: 35 MPa, ignition timing: set for MFB50% at 8 °CA $_{\text{aTDCf}}$

(MFB 90% at 25-30 °CA$_{\text{aTDCf}}$), the emitted light using the high flow injector is on a higher level for a significant number of cycles.

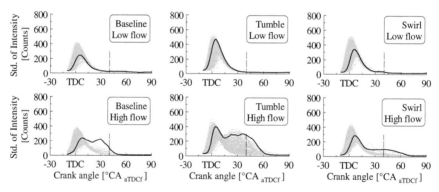

Figure 5.34: Standard deviation of measured intensities of the high-speed imaging at SOI = -300 °CA$_{\text{aTDCf}}$ at 2000 rpm, 0.8 MPa IMEP (WOT), p$_{\text{Inj.}}$: 35 MPa, ignition timing: set for MFB50% at 8 °CA $_{\text{aTDCf}}$

Highlighted are the cycles with the highest std. of intensity at 40 °CA$_{\text{aTDCf}}$ in each graph. For these significant cycles, the high-speed images taken at 40 °CA$_{\text{aTDCf}}$ are plotted in Figure 5.35. For the baseline and tumble configuration in combination with the low flow injector, no diffusive flame is visible. However using the baseline configuration and the high flow injector leads to increased deposit formation on the injector tip and thus to diffusive combustion associated with increased soot emissions. For the optical investigations, the duration of the measurement is too short to build up deposits to a significant extent. Therefore, the PN emissions at the optical investigations are on a lower level ($1\times10^4\,\frac{\#}{cm^3}$) compared to the SOI variation ($3\times10^5\,\#/cm^3$) shown in Figure 5.31 and no significant soot luminosity is measurable. To verify the assumption of the increased tip-sooting effect at this operation point, the engine was operated steady state at $SOI = -300\,°CA_{aTDCf}$ starting with a clean injector. After about 20 minutes of constant operation, the PN emissions increased from $1\times10^4\,\#/cm^3$ to $2\times10^5\,\#/cm^3$.

For the high flow injector, a bright diffusive flame on the piston surface is visible in both the baseline and tumble case (highlighted as a). As the end of combustion is already reached at this time in the cycle, the diffusive flame does not contribute to the power output of the engine, but increases the PN concentration because of the soot formation under the high temperatures and oxygen deficiency. This diffusive combustion on the piston surface is associated with an impingement of liquid fuel on the piston. It is assumed that the tumble motion carried some liquid fuel droplets from the injector to the piston surface on the exhaust side.

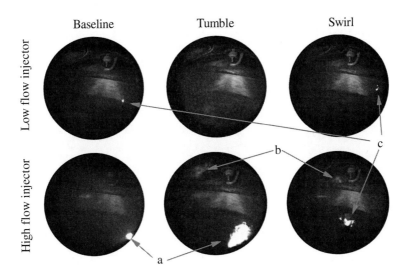

Figure 5.35: Image of high-speed imaging at $40\,°CA_{aTDCf}$ of the Cycles with highest std. of intensity of Figure 5.34 at 2000 rpm, 0.8 MPa IMEP (WOT), $p_{Inj.}$: 35 MPa, ignition timing: set for MFB50% at $8\,°CA_{aTDCf}$

In the baseline case, the light emission is lower and occurred in less cycles (see Figure 5.34), the PN number does not increase as strongly compared to the tumble case. In the tumble case with the high flow injector is a luminous spot close to the injector tip and spark plug (highlighted as b). For the swirl case, a diffusive combustion is obvious for both injectors (highlighted as c). Contrary to the baseline and tumble case with the diffusive combustion on the piston surface, the bright spots seem to be in the gas phase in the swirl configuration. For the low flow injector, two small spots are visible. As of the small size of these spots, detached deposits can be as well assumed to cause this light emission as a diffusive combustion of a fuel-rich zone. In contrast to the liquid fuel on the piston surface, the inhomogeneities in the gas phase are probably caused by insufficient fuel air-mixing. The PN concentrations using the swirl configuration are not significantly increased in this case, as these inhomogeneities occur stochastically and not in a significant number of the 300 cycles measured.

For the operation at WOT with an indicated mean effective pressure of about 0.8 MPa, larger differences between the two injector static flows were measurable. Operating the engine with an optimised injection timing, the PN concentration is on a low level for both injector hydraulic flows. Advancing the injection timing in order to enlarge the time for mixture homogenisation leads to increased PN concentrations. For the low flow injector, an increase of PN emissions is measurable for SOI earlier than -300 °CA $_{aTDCf}$, while the PN concentrations are increased for SOI earlier than -280 °CA $_{aTDCf}$ using the high flow injector. This behaviour was measurable independent of the charge motion configuration, thus indicating difficulties in the mixture formation process. Optical diagnostics showed a significant diffusive combustion on the piston surface on the exhaust side of the engine for all charge motion strategies using the high flow injector.

In combination with the results of the pressure chamber shown in section 4.2, the higher penetration length using the low flow injector does not lead to increased wall impingement at WOT operation. The interaction with the charge motion and the lower droplet diameters are assumed to be more important in terms of mixture preparation.

5.5.3 Influence of injector static flow at boosted operation

The third load point discussed is boosted operation at 2000 rpm. The engine is operated at constant IMEP of 1.4 MPa. Because the engine was prone to knocking at this high engine load, the ignition timing was delayed to prevent the engine from damage. The MFB50% was therefore kept constant at 22 °CA$_{aTDCf}$, while the air-fuel ratio was kept constant at stoichiometry ($\lambda = 1$). The boost pressure was adapted to meet the engine load as well as the air-fuel ratio and was set about 20 to 35 kPa above ambience. The injection pressure used with both injectors was 35 MPa. Therefore, the injection timing for the low flow injector needed to be prolonged to inject the same amount of fuel. The valve timings for the investigations were set to a slight negative valve overlap. Therefore, no effects of scavenging should influence the results.

Emission characteristics

Again SOI variations were performed with both injectors and the three charge motion strategies. The resulting PN concentrations are shown in Figure 5.36. Compared to the WOT operation in Figure 5.31, the PN concentrations for both injectors are on a higher level using the baseline configuration. Like at WOT operation, the reduction of the static flow only shows a significant effect at early injection timings. For all charge motion strategies, the low flow injector shows significantly lower PN concentrations at early injection timings ($> -300\,°CA_{aTDCf}$). This is assumed to be caused by the higher flow rate and thus higher penetration at earlier timings by the high flow injector. Therefore, an impingement of liquid fuel causing poolfire is more likely. This assumption is supported by the higher hydrocarbon concentration for the early injection timings and using the high flow injector, as shown in Figure 5.37.

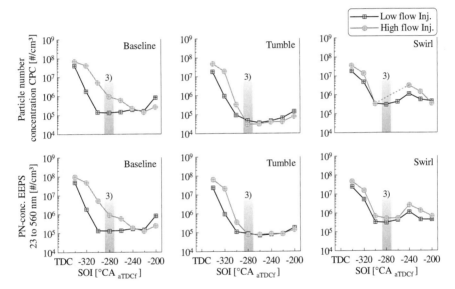

Figure 5.36: PN emissions for the SOI variations with high flow and low flow injector at 2000 rpm, 1.4 MPa IMEP, $p_{Inj.}$: 35 MPa, ignition timing: set for MFB50% at $22\,°CA_{aTDCf}$

Using tumble charge motion, the PN concentrations can be reduced independent of the static flow of the injector. The tumble motion leads to a better mixture formation and thus lower CoV of IMEP for all injection timings, shown in Figure 5.37. The CoV of IMEP values also suggest that the higher PN concentrations for both injectors with late SOI using the baseline configuration are caused by insufficient mixture preparation. However, for the swirl configuration the PN concentrations are higher compared to the tumble configuration and on a similar level compared to the baseline configuration.

Compared to the baseline and the tumble case, the increased PN emissions for the SOI at -240 °CA$_{aTDCf}$ is noticeable (highlighted as 4) for the swirl case. As described in section 5.4.2 on page 96, these increased PN emissions were caused by a spray-valve interaction. To get more information on the influence of the injector static flow at boosted operation without the influence of the spray-valve interaction, high-speed imaging measurements were performed for SOI = -240 °CA$_{aTDCf}$ (highlighted as 3). Using the swirl inlay, one intake port is isolated from the charge air flow. The increased PN emissions were caused by the generation of the large scale charge motion. For this injection timing, a strong spray-valve interaction was visible. Because there was no charge air flow at one of the intake valves, the impinged fuel on the intake valve did not take part in the mixture formation process. This caused inhomogeneities in the gas phase and thus a diffusive combustion with increased soot formation. Therefore, the CoV of IMEP is increased as well as the PN concentration in the exhaust gas, as shown in Figures 5.36 and 5.37.

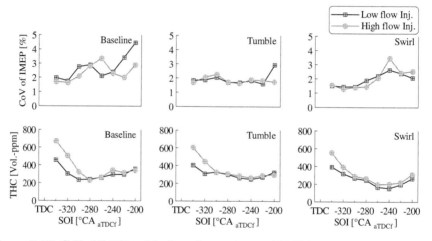

Figure 5.37: CoV of IMEP and hydrocarbon emissions (THC) for the SOI variations with high flow and low flow injector at 2000 rpm, 1.4 MPa IMEP, p$_{Inj.}$: 35 MPa, ignition timing: set for MFB50% at 22 °CA$_{aTDCf}$

Analysis at injection timing -280 °CA$_{aTDCf}$

To get a better understanding of the results, additional data of the high-speed imaging for a start of injection at -280 °CA$_{aTDCf}$ (highlighted as 3 in Figure 5.36) are shown in Figure 5.38. The std. of intensity at 45 °CA$_{aTDCf}$ is lower for all cases and does not show as strong outliers as at WOT operation (shown in Figure 5.34). Highlighted are the cycles with the highest std. of intensity at 45 °CA$_{aTDCf}$ in Figure 5.38. Only for the cases baseline and swirl using the high flow injector, slightly increased values in single cycles are obvious. This indicates that the increased PN emissions are not caused by a fuel impingement on a component. An insufficient mixture formation process is more likely.

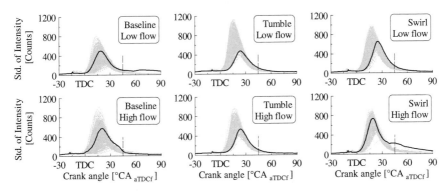

Figure 5.38: Standard deviation of measured intensities of the high-speed imaging at SOI = -280 °CA$_{aTDCf}$ at 2000 rpm, 1.4 MPa IMEP, p$_{Inj.}$: 35 MPa, ignition timing: set for MFB50% at 22 °CA$_{aTDCf}$

In Figure 5.39, the high-speed images taken at 45 °CA$_{aTDCf}$ of the highlighted cycles from Figure 5.38 are plotted. For the tumble configuration with the lowest PN concentrations measured, no luminous spots are detectable in the combustion chamber. However, the overall light emission of the cycle shown is on a higher level compared to the other configurations. This is caused by a late combustion phasing of these single cycles and not contributed to soot luminosity.

For the baseline case, the high flow injector emitted a higher PN concentration compared to the low flow injector. Taking a look at the cycle with the highest luminosity of the high-speed imaging, bright spots in the gas phase are visible using the low flow injector (highlighted as α in Figure 5.39). The high flow injector shows a diffusive combustion at the piston surface (highlighted as β in Figure 5.39). For the tumble configuration, only a slight diffusive combustion at the injector tip is visible using the low flow injector (highlighted as γ in Figure 5.39). For the swirl configuration with the highest PN concentrations at SOI = -280 °CA$_{aTDCf}$, diffusive flames in the gas phase are visible using the high flow injector (highlighted as α in Figure 5.39).

As it is shown in Figure 5.38, these events are stochastically distributed, leading to the assumption, that several cycles with insufficient mixture preparation increase the average PN concentration in the exhaust gas. With the currently available particle measurement devices, the maximum measurement frequency is limited to 10 Hz. Additionally, the long pathways to the measurement device and the sample preparation in the VPR lead to mixing of consecutive combustion cycles. Therefore, a separation of the PN emissions of single cycles is not easily possible. Measurements by Cudeiro Torruella et al. [29] using Laser induced incandescence (LII) support the assumption, that under some operating conditions, single cycles with high soot emissions cause the major part of the overall soot emissions.

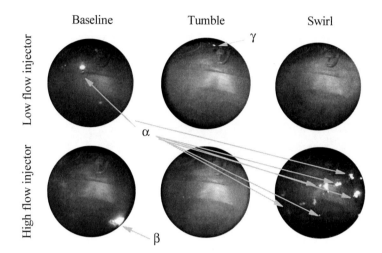

Figure 5.39: High-speed images at $45\,°\mathrm{CA}_{\mathrm{aTDCf}}$ of the cycles with highest std. of intensity of Figure 5.38 at 2000 rpm, 1.4 MPa IMEP, $\mathrm{p}_{\mathrm{Inj.}}$: 35 MPa, ignition timing: set for MFB50% at $22\,°\mathrm{CA}_{\mathrm{aTDCf}}$

For the operation at high engine load of 1.4 MPa IMEP, different PN emissions characteristics between the two injector hydraulic flows are measurable for the baseline and swirl configuration. With tumble charge motion, the two injectors show comparable emission behaviour on a low level. For the baseline configuration, the increased PN emissions using the high flow injector could be attributed to fuel impingement on the piston on the exhaust side, comparable to the WOT operation in section 5.5.2. For the swirl case, stochastic inhomogeneities in the gas phase are assumed to cause the higher PN emissions using the high flow injector.

In terms of the hydraulic flow of the injector, the negative influence of a prolongued injection time for the same injection pressure and fuel mass does not overpower the positive aspects of the reduced nozzle hole diameters, such as the reduced SMD.

5.5.4 Influence of injection pressure at boosted operation

By reducing the static flow of the injector, either the injection duration needs to be prolonged or the injection pressure needs to be increased to keep the injected fuel mass constant. The effect of the prolonged injection duration was shown in the previous sections and the effect of an increased injection pressure is discussed in the following part. A proportion of these results was published in [18].

Once again the engine was operated at 2000 rpm and with boosted operation (1.4 MPa IMEP). As the engine was prone to knocking at this high engine load, the ignition timing was delayed to save the engine from damage. The MFB 50% was therefore kept constant at 22 °CA$_{aTDCf}$, while the air-fuel ratio was kept constant at stoichiometry ($\lambda = 1$). The boost pressure was adapted to meet the engine load and the air-fuel ratio and was about 200 to 350 mbar above ambient. The valve timings for the investigations were set to a slight negative valve overlap. Therefore, no effects of scavenging should influence the results.

Emission characteristics

The injection pressure can be increased up to 50 MPa by using the low flow injector. The following graphs show the results of SOI variations using 20, 35 and 50 MPa. In Figure 5.40, the PN concentrations measured are shown.

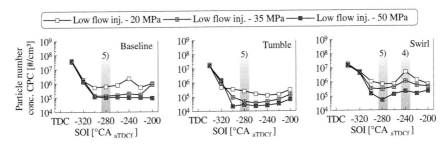

Figure 5.40: PN emissions for the SOI variations with different injection pressure at 2000 rpm, 1.4 MPa IMEP, low flow injector, ignition timing: set for MFB50% at 22 °CA$_{aTDCf}$

For all three charge motion strategies, a significant reduction of the PN concentration is measurable by increasing the injection pressure up to 35 MPa. The only exception are the early injection timings at -320 °CA$_{aTDCf}$ and -340 °CA$_{aTDCf}$ with the piston impingement. An increased injection pressure does not lead to decreased PN concentrations in this case. By further increasing the injection pressure up to 50 MPa, an additional reduction of the PN concentrations is obvious. Even for late injection timings, the PN concentrations stay on a low level for all charge motion strategies. This can be caused by the prolongued time for mixture preparation due to the higher injection pressure and thus higher mass flow rate. Additionally the smaller SMD leads to a bigger surface-volume ratio and thus faster evaporation. This also leads to higher engine running smoothness, shown by the lower values of CoV of IMEP in Figure 5.41. The hydrocarbon emissions are not significantly influenced by the higher injection pressure.

As already discussed for Figure 5.36, an increased PN concentration is measurable for the start of injection at -240 °CA$_{aTDCf}$. This is caused by the generation of the swirl by covering one intake port. This effect can be reduced significantly by increasing the injection pressure (highlighted as 4). The smaller droplets with higher impulse and lower surface-volume ratio seem to support the mixture formation process more strongly than the interaction of spray and intake valve interfere the process.

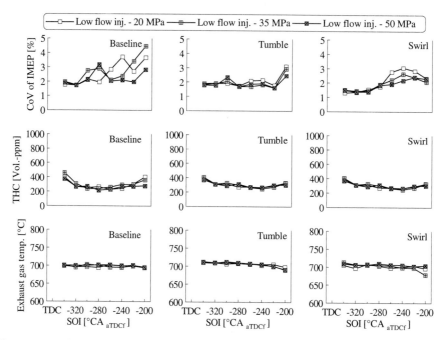

Figure 5.41: CoV of IMEP and hydrocarbon emissions (THC) for the SOI variations with different injection pressure at 2000 rpm, 1.4 MPa IMEP, low flow injector, ignition timing: set for MFB50% at 22 °CA $_{aTDCf}$

Analysis at injection timing -280 °CA$_{aTDCf}$

Once again the particle size distributions at SOI = -280 °CA$_{aTDCf}$ (highlighted as 5) in Figure 5.40) for the three injection pressures and all charge motion strategies are shown in Figure 5.42. For all charge motion strategies, the amount of nucleation mode particles is not significantly influenced by the increased injection pressure. However, the reduction of accumulation mode particles by the increased injection pressure is obvious for all cases. This supports the assumption that the reduced particle emission with higher injection pressure is caused by an improved evaporation process and therefore a faster mixture formation process. This indicates that the reduction of accumulation mode particles is caused by reduced soot emissions.

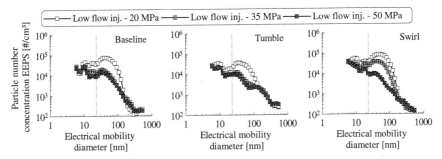

Figure 5.42: Particle size distributions for SOI = -280 °CA$_{aTDCf}$ with different injection pressure at 2000 rpm, 1.4 MPa IMEP, low flow injector, ignition timing: set for MFB50% at 22 °CA $_{aTDCf}$

To validate this assumption, measurements using the high-speed imaging setup were performed for 20 MPa, 35 MPa and 50 MPa. The calculated std. of intensity is plotted versus the crank angle for the three charge motion strategies in Figure 5.43. As already shown in the last section, no significant diffusive combustion cycles are visible during the measured 300 cycles. This shows the challenges to measure soot emissions in GDI engines. If the engine operates at low soot levels, the cause for these low soot emissions is not easy to identify.

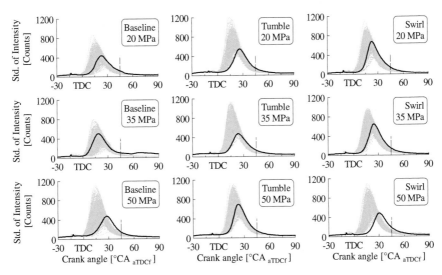

Figure 5.43: Standard deviation of measured intensities of the high-speed imaging at SOI = -280 °CA$_{aTDCf}$ with different injection pressure at 2000 rpm, 1.4 MPa IMEP, low flow injector, ignition timing: set for MFB50% at 22 °CA $_{aTDCf}$

Nevertheless, the images at 45 °CA$_{aTDCf}$ of the cycles with highest std. of intensity of Figure 5.43 are shown in Figure 5.44. For operation with 20 MPa injection pressure, some luminous spots are detectable, highlighted as I. Increasing the injection pressure up to 50 MPa reduces the single cycles with insufficient mixture preparation. During the 300 cycles measured, no cycles with higher luminosity were measured.

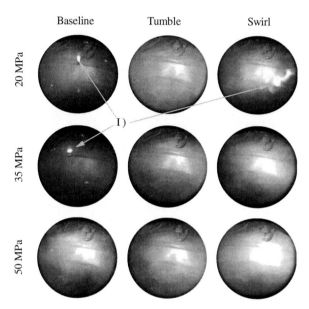

Figure 5.44: High-speed images at 45 °CA$_{aTDCf}$ for SOI = -280 °CA$_{aTDCf}$ with different injection pressure at 2000 rpm, 1.4 MPa IMEP, low flow injector, ignition timing: set for MFB50% at 22 °CA $_{aTDCf}$

5.6 Influence of fuel composition and fuel temperature on PN formation

As discussed in section 2.4.6, the fuel composition and condition have significant influence on both mixture formation and emission formation. However, recent investigations focussed on low engine load (Aikawa et al. [2], Leach et al. [116]) and catalyst heating operation (Dageförde et al. [32]). In this section, results of changed fuel conditions and changed fuel compositions are shown for high engine load.

5.6.1 Fuel temperature

The fuel conditions, such as density and viscosity, change depending on the temperature and pressure. As already shown in section 5.5.4, the fuel pressure shows significant influence on both the mixture formation and particle emissions. To show the influence of the fuel temperature, several variations with heated fuel were performed. Thereby the fuel was heated from baseline conditions[9] with 30°C to 45°C and 60°C. Summarised, the fuel temperature did not affect the combustion process and the emission formation for steady-state operation at high engine load. For low engine load with throttled operation and thus pressure levels below ambient inside the cylinder, the fuel temperature shows significant impact on mixture formation with phenomena like flash boiling as indicated for instance by Weber and Leick [198] and Kamoun et al.[97]. However, the focus of this research work was on high engine load. There was an effect of the fuel temperature measurable concerning the deposit formation at the injector tip. As shown in section 2.4.4, the deposit formation at the injector tip can increase the PN emissions by about one order of magnitude. To show the fuel temperature dependency on the deposit formation, the engine was operated in a steady state at 2000 rpm and an indicated mean effective pressure of 1.4 MPa with an injection pressure of 20 MPa. As shown in Figure 5.45, a steep increase of PN emissions with time was measurable for all three fuel temperatures. However, the increased fuel temperatures showed potential to decelerate the deposit formation. The formation of deposits at the injector tip does not influence the mixture formation nor the combustion in these investigations, which accords well to the literature results discussed in section 2.4.4. Thus neither the emission of hydrocarbons nor the combustion stability are affected.

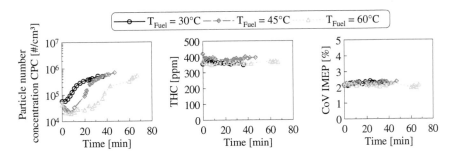

Figure 5.45: Influence of fuel temperature on deposit formation at injector tip for constant operation 2000 rpm, 1.4 MPa IMEP, low flow injector, $p_{Inj.}$: 20 MPa, ignition timing: set for MFB50% at 22 °CA aTDCf

---

[9]the fuel temperature was measured in the fuel line close to the injector

5.6.2 Fuel composition

For the further reduction of PN emissions, Alkylate fuel was used in comparison to RON 95. As Alkylate fuel does not contain aromatic compounds, the soot formation process should be reduced according to the soot formation theory as shown in section 2.2. Thereby the vapour pressure, as the second contributor to the PM Index by Aikawa et al. [2], is comparable to RON 95. Dageförde et al. already showed the potential to reduce PN emissions at catalyst heating operation. However, both for catalyst heating operation and the investigations by Aikawa et al. [2] and Leach et al. [116] the baseline PN emissions were on a comparatively high level.

In this work, the engine was operated at 2000 rpm and 1.4 MPa indicated mean effective pressure. Thereby the high flow injector was used with 30 MPa injection pressure and the ignition timing was set to knock-limited spark advance. The resulting PN emissions of a SOI variation using the three charge motion strategies baseline, tumble and swirl with both RON 95 and Alkylate fuel are shown in the upper graphs of Figure 5.46. These results were partially published in [16].

The previously discussed characteristic PN emission for GDI homogeneous engines is obvious for both fuels. Thereby increased PN emissions are measurable for early and late injection timings. Early injection leads to a liquid fuel film on the piston causing poolfire, while a late injection leads to a reduced time for mixture homogenisation and thus inhomogeneities in the gas phase. For the baseline configuration the fuel shows a strong influence on PN emissions. For all SOI except the latest, a reduction of PN by about one order of magnitude is possible when using Alkylate fuel.

As shown before, the emission of PN using RON 95 is overall the lowest with tumble inlay. Thereby the Alkylate fuel shows the lowest impact on PN emission. Only at SOI timings with high emission of PN (early and late injection), a significant reduction of PN is measurable. Except for the early injection, the PN emission is nearly independent of the SOI using tumble and RON 95. This low affect of the Alkylate fuel is linked to the knocking behaviour of Alkylate fuel caused by the lower sensitivity compared to RON 95. Li, Ghazi and Sohrabi showed that for knocking operation in a SI hydrogen engine an increased oil consumption was measurable [120]. Increasing oil consumption leads to local zones with increased hydrocarbon fractions and thus to a local combustion with oxygen deficiency leading to higher PN emissions.

The PN emissions using swirl configuration show a behaviour comparable to the baseline configuration. Using RON 95, the emission is on a higher level compared to the tumble case. With Alkylate fuel, the PN emissions drop down by about one decade. As already discussed in section 5.4.1, the increase of PN emissions using the swirl configuration for SOI about -240 °CA $_{aTDCf}$ is attributable to spray valve interactions and obviozus for both fuels.

For most of the operations a minimum PN emission was measured with an SOI of -260 °CA $_{aTDCf}$. The particle size distribution for these measurements is shown in the lower graphs of Figure 5.46. For all the charge motion configurations, the nucleation mode as well as the accumulation mode are on a lower level using Alkylate fuel. The higher PN emission using tumble and Alkylate fuel, compared to the other charge motion configurations with Alkylate fuel is driven by a higher amount of accumulation mode particles. As discussed in section 2.1, the nucleation mode most likely consists of nanoparticles formed from volatile precursors while the accumulation mode mainly consists of larger carbonaceous agglomerates that have survived the combustion process.

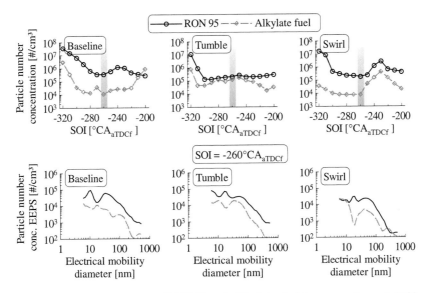

Figure 5.46: PN emissions using RON 95 and Alkylate fuel in comparison at 2000 rpm, 1.4 MPa IMEP, high flow injector, $p_{Inj.}$: 30 MPa, ignition timing: knock-limited spark advance

The resulting emissions of hydrocarbon and carbon dioxide as well as the CoV of IMEP and the centre of combustion (MFB50%) are shown in Figure 5.47. For most of the injection timings, the THC emissions are slightly lower using Alkylate fuel. The increase of THC emissions for early injection timings is attributed to fuel impingement on the piston. For the swirl and the tumble configuration, the engine was more prone to knocking using Alkylate fuel, resulting in a later MFB50%. For the baseline configuration, the MFB50% is on the same delayed timing of about 28 °CA $_{aTDCf}$ for both fuels. The cyclic variations depend directly on the MFB50%. For late combustion phasing, the cyclic variations increase. The emission of carbon dioxide is lower for all configurations and timings because of the lower C/H-ratio of Alkylate fuel.

There are two explanations for Alkylate fuel being more prone to knocking compared to RON 95 (see Table 3.2). One explanation is the lower evaporation heat (see Table 3.2) and thus reduced charge cooling effect. The temperature at the end of compression and thus at ignition timing is higher and increases the self-ignition processes. The second explanation is the lower sensitivity[10] of Alkylate fuel. Measurements by Kalghatgi [94, 95, 96], Amer et al. [6] as well as results of Mittal and Heywood [135, 136] show that fuels with a lower sensitivity can cause a lower knocking resistance especially at turbocharged conditions.

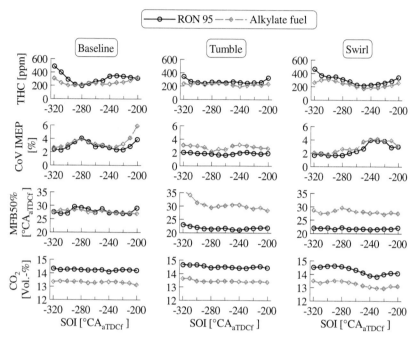

Figure 5.47: Combustion and emissions using RON 95 and Alkylate fuel in comparison at 2000 rpm, 1.4 MPa IMEP, high flow injector, $p_{Inj.}$: 30 MPa, ignition timing: knock-limited spark advance

Additional measurements with the high-speed camera using the baseline configuration at SOI = -260 °CA $_{aTDCf}$ were made to explain the lower PN emissions using Alkylate fuel. As the physical properties of RON 95 and Alkylate fuel are close together, there was no measurable effect on the injection process.

However, concerning the deposit formation, a tendency to reduce the deposit formation at the injector tip was measurable. In Figure 5.48, the standard deviation of the measured intensities for both fuels are plotted versus the crank angle. As the combustion phasing was identical, neither differences in maximum intensity nor cyclic variations are obvious between the two fuels. Taking a closer look at the end of combustion gives a hint on the increased PN emissions using RON 95, as shown in the lower graph in Figure 5.48. It is obvious that the intensity for the cycles using Alkylate fuel drops

---

[10]Sensitivity = RON - MON

down earlier. The higher intensities using RON 95 can be attributed to the diffusive combustion at the injector tip. Therefore, more PN is emitted as a result of diffusive combustion.

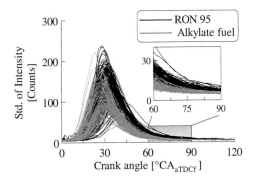

Figure 5.48: High-speed measurements for the comparison of RON 95 and Alkylate fuel using the baseline configuration at 2000 rpm, 1.4 MPa IMEP, high flow injector, $p_{Inj.}$: 30 MPa, ignition timing: knock-limited spark advance

Additionally the deposit formation was investigated using a spectrograph in [16]. A calibration lamp was used to determine the reduction of transmission by deposits on the optical accesses of the engine. It could be demonstrated that in contrast to RON 95, Alkylate fuel did not significantly build up deposits on the optical accesses. These results support the assumption that Alkylate fuel has the potential to reduce PN emissions of GDI engines by the reduced tendency to build up deposits and thus reduce tip-sooting.

Summed up, Alkylate fuel shows potential to reduce PN emissions from GDI engines. As the physical properties are comparable to RON 95, there is no need to change the engine settings for the use of Alkylate fuel. However, at high engine load and boosted operation, the lower sensitivity and lower evaporation heat of Alkylate fuel leads to increased engine knocking, resulting in a worse combustion phasing and thus lower efficiency.

5.7 Influence of lubricant and coolant temperature on PN formation

The temperature of lubricant and cooling influence the temperature inside the cylinder. As shown by Heywood, the cooling temperature influences the temperature of the cylinder head, the liner, piston and thus via heat transfer of the charge in the combustion chamber directly [79]. The temperature of the charge affects the evaporation of the liquid fuel and thus the time for the mixture formation process. Further the oil temperature influences the temperature of the piston if a piston spray nozzle is equipped to the engine. The temperature of the piston is assumed to influence the charge temperature too. Additionally, the evaporation process of impinged fuel on the piston surface is influenced by the oil temperature.

Investigations at high engine load of 1.4 MPa IMEP at 2000 rpm were conducted to show the effect of reduced temperatures of both cooling and lubricant.

5.7.1 Engine lubricant temperature

The engine lubricant has the function to reduce the frictional resistance of the engine to a minimum in order to ensure maximum mechanical efficiency, to protect the engine against wear and to cool the piston and regions where friction work is dissipated [170]. To show the influence of the lubricant temperature on the PN emissions, a SOI variation was performed for a lubricant temperature of 90 °C as baseline, 60 °C and 45 °C using the baseline charge motion configuration. The injection pressure for the low flow injector was set to 20 MPa, the ignition timing was set to a constant MFB50% at 22 °C aTDCf and the coolant temperature to 90 °C. The resulting PN emissions as well as the hydrocarbon emissions and the combustion stability (CoV IMEP) are shown in Figure 5.49. It is obvious that neither the PN nor the THC emissions are influenced by the temperature of the lubricant. Especially at early injection timings, with the related fuel impingement on the piston, a negative influence of cooler lubricant on the PN emissions was not measurable. This indicates that the temperature of the piston is more strongly influenced by the temperature of the cooling. The increased PN emissions at SOI = -240 °C aTDCf were caused by a spray-valve interaction as described in section 5.4.2.

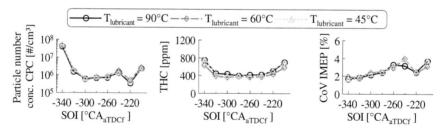

Figure 5.49: Effect of changing lubricant temperature on PN formation and exhaust gas emissions at 2000 rpm, 1.4 MPa IMEP, low flow injector, $p_{Inj.}$: 20 MPa, ignition timing: set to MFB50% at 22 °C aTDCf

In Figure 5.50, the particle size distribution for SOI = -200 °C aTDCf (left), SOI = -260 °C aTDCf (middle) and SOI = -300 °C aTDCf (right) are shown. These size distributions show the low emissions for the earlier injection timings and the increased

PN emissions for late injection timing (SOI = -200 °CA $_{aTDCf}$) caused by a too short period of time for mixture preparation. However, no clear influence of the lubricant temperature on the PN size distribution is measurable. The increased PN emission is obvious for 60 °C at SOI = -200 °CA $_{aTDCf}$. However, for the further reduction of the lubricant temperature to 45 °C, the emission is lower for the same settings. Therefore, it is assumed that the higher PN emissions with 60 °C lubricant temperature were caused by the high cyclic variations at these operating conditions.

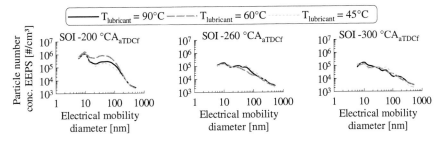

Figure 5.50: Effect of changing lubricant temperature on PN formation and exhaust gas emissions at 2000 rpm, 1.4 MPa IMEP, low flow injector, $p_{Inj.}$: 20 MPa, ignition timing: set to MFB50% at 22 °CA $_{aTDCf}$

The lubricant temperature is assumed to have a stronger influence on PN emissions at acceleration processes, as shown by Köpple et al. [110, 111]. Thereby a temporary reduction of the piston cooling, in order to increase the piston surface temperature faster, showed benefits compared to operation with regular piston cooling.

5.7.2 Engine coolant temperature

The influence of the coolant temperature was investigated comparable to the influence of the lubricant temperature. The engine speed was again set to 2000 rpm and the engine load to 1.4 MPa IMEP with the ignition timing set for an MFB50% of 22 °CA $_{aTDCf}$. The injection pressure for the low flow injector was set to 20 MPa and the lubricant temperature to 90 °C. These results were partially published in [15].
In Figure 5.51 the resulting PN and hydrocarbon emissions are shown with the cyclic variations of the SOI variations for the different coolant temperatures. A clear tendency for PN and hydrocarbon emissions to increase for decreasing temperature is obvious, whereas the cyclic variations are on the same level. Reducing the coolant temperature to 60 °C, the PN emissions for the optimised injection timing of SOI = -300 °CA $_{aTDCf}$ are on the same level. However, for earlier and later injection timings, there is a stronger increase of the PN emissions compared to the 90 °C coolant temperature. It is assumed that the reduced coolant temperature reduces the charge air temperature and thus decelerates the evaporation process, leading to inhomogeneities

in the gas phase for late injection timings. For the further reduction of the coolant temperature to 30 °C, the PN emissions are increased for all injection timings.

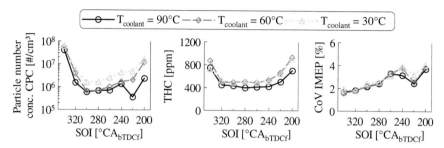

Figure 5.51: Resulting PN and hydrocarbon emissions as well as cyclic variations for the different coolant temperatures at 2000 rpm, 1.4 MPa IMEP, low flow injector, p_{Inj}: 20 MPa, ignition timing: set to MFB50% at 22 °C aTDCf

The particle size distributions at SOI -220 °CA aTDCf and -300 °CA aTDCf for the different coolant temperatures are plotted in Figure 5.52. For SOI = -300 °CA aTDCf, the particle size distributions for the three coolant temperatures are close together. The amount of nucleation mode particles is comparable, while the amount of accumulation mode particles is increased for the colder cases. For the late injection timing of SOI = -220 °CA aTDCf, both the nucleation mode and the accumulation mode particles are increased. The increase of nucleation mode particles thereby supports the assumption of mixture inhomogeneities caused by the lower in-cylinder temperatures with lower coolant temperatures.

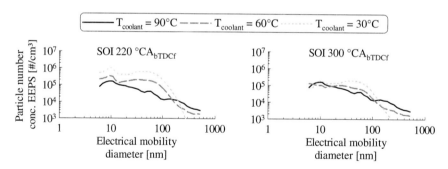

Figure 5.52: Particle size distributions measured for SOI = -220 °CA aTDCf and SOI = -300 °CA aTDCf for the different coolant temperatures at 2000 rpm, 1.4 MPa IMEP, low flow injector, p_{Inj}: 20 MPa, ignition timing: set to MFB50% at 22 °C aTDCf

The in-cylinder temperatures are important for the mixture formation process and assumed to cause inhomogeneities in the mixture for lower coolant temperatures. As the in-cylinder gas temperature is difficult to measure, the temperatures of the gas were calculated using the two-zone TPA model as presented in section 3.3. In the left graph of Figure 5.53, the calculated temperatures during the injection process for SOI = -200 °CA aTDCf are shown, while the temperature of the unburned zone is shown in the graphs on the right hand side. The assumed reduced in-cylinder temperatures are obvious for the reduced coolant temperatures. For the 30 °C coolant temperature, the gas temperature in the cylinder at start of injection is about 8 K lower compared to 90 °C coolant temperature conditions. The reduced coolant temperature also leads to reduced temperatures during the compression stroke, indicating that the evaporation process is decelerated due to the reduced temperature difference of liquid and gas phase.

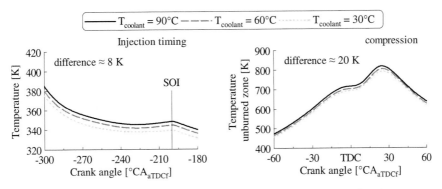

Figure 5.53: Calculated in-cylinder temperatures for the different coolant temperatures at SOI = -200 °CA aTDCf, 2000 rpm, 1.4 MPa IMEP, low flow injector, $p_{Inj.}$: 20 MPa, ignition timing: set to MFB50% at 22 °CA aTDCf

Optical diagnostic techniques were further on applied to analyse the particle formation process. Firstly, the high-speed camera was used for the two coolant temperatures 30 °C and 90 °C. Secondly, the spark plug with fibre optical access was used to determine the influence of reduced coolant temperatures on the PN emissions.

The standard deviation of intensity of the 80 cycles measured are shown in Figure 5.54, with 90 °C coolant temperature in the left hand graph and 30 °C in the right hand graph. As shown in the previous section, using the low flow injector under hot, steady-state conditions, the cyclic variations are low. However, one cycle shows higher light emission, highlighted as cycle 06. For the reduced coolant temperature, the cyclic variations of the emitted light at late crank angle (>30 °CA aTDCf) are higher, indicating a higher amount of diffusive combustion. Even though the cyclic variations are high, one cycle appears to emit significantly more light compared to the others, highlighted as cycle 68.

The images of the cycles highlighted in Figure 5.54 from 15 °CA aTDCf on in steps of 5 °CA are shown in Figure 5.55. For the 90 °C case, a mixture inhomogeneity in the

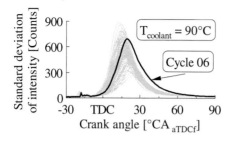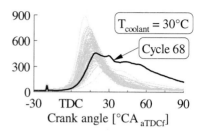

Figure 5.54: Standard deviation of intensity of high-speed images for the different coolant temperatures for SOI = -300 °CA aTDCf at 2000 rpm, 1.4 MPa IMEP, low flow injector, $p_{Inj.}$: 20 MPa, ignition timing: set to MFB50% at 22 °CA aTDCf

gas phase, causing a diffusive combustion, is visible. Tip-sooting is also visible at the injector (top, centre). A bright diffusive combustion is located in the centre of the combustion chamber and moves from the centre to the exhaust side. Caused by the downwards moving piston, the diffusive combustion gets carried downwards for later degree crank angle. The PN emissions for operation with 20 MPa injection pressure and baseline configuration are on slightly increased level (about $6 \cdot 10^6$ #/cm³) compared to the higher injection pressures, as shown in section 5.5.4. These increase in PN emissions is assumed to be caused by insufficient mixture preparation causing fuel rich zones in the gas phase and thus a diffusive combustion, as shown in the upper pictures in Figure 5.55.

For the 30 °C coolant temperature case, more cycles showed high light emissions after the regular combustion process, indicating diffusive combustion. One cycle with extraordinary high light emissions (cycle 68) was picked and is shown in the lower picture row of Figure 5.55. Two luminous spots are identifiable: Severe tip-sooting in the top centre of the picture and a lot of luminous spots in the centre of the combustion chamber. As the temperature in the combustion chamber is reduced due to the lower coolant temperature, the mixture formation process is considerably more difficult.

The second optical diagnostic technique used to determine the influence of reduced coolant temperatures on the combustion process was the spark plug with fibre optical access. The soot integral was determined as described in section 3.4.4. As the spark plug with fibre optical access is easily applicable and does not need to be cleaned for each measurement, it was possible to perform all SOI variations with the three coolant temperatures, shown in Figure 5.51. The soot integral is shown in the graphs of Figure 5.56. The calculated soot integral of the 500 cycles measured is plotted on the vertical axis of the graphs with the end of combustion (MFB90%) of the cycles on the horizontal axis. The results for SOI = -240 °CA aTDCf (upper row), SOI = -280 °CA aTDCf (middle row) and SOI = -340 °CA aTDCf (bottom row) are plotted for coolant temperatures 90 °C (left hand column), 60 °C (middle column) and 30 °C (right hand column). It should be noted that almost all cycles of the variations (500 cycles for each graph) show an MFB90% prior to 60 °CA aTDCf, which is the start

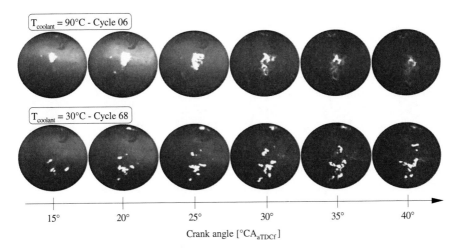

Figure 5.55: Selected cycles with highest standard deviation of intensity at 60 °CA aTDCf of Figure 5.54 at 2000 rpm, 1.4 MPa IMEP, low flow injector, $p_{Inj.}$: 20 MPa, ignition timing: set to MFB50% at 22 °CA aTDCf

of integration for the soot integral.

First thing to mention are the increased cyclic variations of MFB90% for lower coolant temperatures and late injection timing (SOI = -240 °CA aTDCf). Noticeable are the high levels of soot integral for these cycles with late combustion phasing. For the 90 °C coolant temperature, the cycles with an MFB90% prior to 30 °CA aTDCf show soot integrals with values below 3, which is a sign of a clean combustion without any diffusive flame. For the combustion phasing between 30 °CA aTDCf and 50 °CA aTDCf a few cycles with soot integrals from 3 to 5 were measured, indicating a combustion with diffusive combustion past the regular combustion. Only for one cycle with the latest combustion phasing measured of about 50 °CA aTDCf, a soot integral with a value of about 9.5 was measured, indicating a severe diffusive combustion. These results support the assumption, that the increased PN emissions for the late injection timing for hot, steady-state engine operation are caused by mixture inhomogeneities with a stochastic character.

For the reduced coolant temperature, the number of cycles with late combustion phasing increases significantly. A higher number of cycles with increased soot integral goes along with the larger number of cycles with retarded combustion phasing. It is notable that combustion cycles with relatively early combustion phasing but increased soot integral appear for lower coolant temperatures. It is assumed that these cycles with increased soot integral show weak diffusive combustion and thus increase the PN emissions. For the optimised injection timing of SOI = -280 °CA aTDCf, the cyclic variations of MFB90% are on a lower level. For the 90 °C and the 60 °C coolant temperature, both the particle number concentrations and the soot integrals of the cycles are on a low level compared to the retarded injection timing (upper row). For the 30 °C coolant temperature, the PN emissions at SOI = -280 °CA $_{\text{aTDCf}}$ are lower compared to SOI = -240 °CA $_{\text{aTDCf}}$, but on a higher level compared to the higher coolant temperatures. The soot integral for a large number of cycles with relatively early MFB90% is thereby above 5 (grey marked square), indicating cycles with diffusive combustion.

For the advanced injection timing of -340 °CA $_{\text{aTDCf}}$ the PN emissions are increased significantly due to fuel impingement on the piston and thus a diffusive combustion on the piston surface. However, a trend of increased PN emissions for reduced coolant temperatures is obvious, which is combined with increased values of the soot integral for the same MFB90% values. It is assumed that the piston surface temperature is reduced with decreased coolant temperatures and thus the evaporation rate of the liquid fuel is lowerd resulting in more pronounced diffusive combustion.

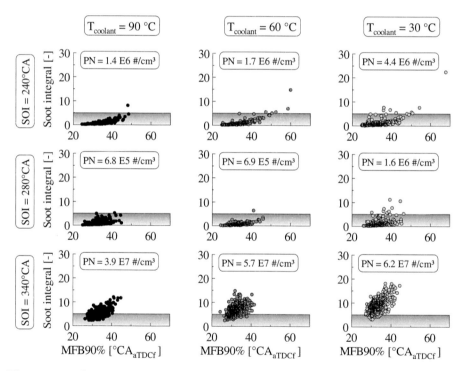

Figure 5.56: Soot integrals calculated with the signals of the spark plug with fibre optical access for the different coolant temperatures using the baseline configuration at 2000 rpm, 1.4 MPa IMEP, low flow injector, $p_{\text{Inj.}}$: 20 MPa, ignition timing: set to MFB50% at 22 °CA $_{\text{aTDCf}}$

5.8 Inflammation and PN formation

To show the influence of the inflammation process on the PN formation of the engine, a high-frequency ignition system (HFI) was compared to the baseline transistor-coil ignition system (TCI). As shown by different publications, the HFI system has the potential to reduce the cyclic variations of gasoline engines [73, 74, 75, 77, 148]. Wolf et al. thereby state that due to a voluminous inflammation area as well as the radical-chemical inflammation rather than the radical-thermal process of conventional ignition systems, the HFI allows a stable inflammation with low cyclic variations [206]. As shown in the previous sections, increased cyclic variations tend to increase the PN emissions. A reduction of the cyclic variations without changing the combustion chamber or fuel characteristics therefore is assumed to have the potential to reduce PN emissions.

To demonstrate the potential of enhanced inflammation process, the HFI system shown in section 3.2.1 was compared to the conventional TCI system used in the previous sections. The operation points concerning the residual gas concentration and already discussed in section 5.3, are therefore compared to the operation with the HFI system and same engine parameters. In this section, only the results using the supercharger configuration without exhaust gas back-pressure are considered. The engine was operated at an engine speed of 2000 rpm and an IMEP of 1.4 MPa. The high flow injector was used with an injection pressure of 20 MPa. The spark timing was set to knock-limited spark advance. These measurements were also the basis for the publications [19] and [74].

To compare the ignition systems under different conditions, the exhaust valve open timing was varied using the three charge motion strategies previously discussed (baseline, tumble, swirl). The measured PN concentrations are shown in Figure 5.57. As previously shown, the implementation of a large scale charge motion reduces the PN emissions for all exhaust valve open timings. Caused by the low exhaust gas back-pressure, the amount of residual gas is relatively low, as shown in Figure 5.14. However, for all charge motion configurations and valve timings, the PN emissions can be reduced by implementing a high-frequency ignition system and thus enhancing the inflammation process.

To get more information on the composition of the PN emissions, the particle size distributions for the earliest exhaust valve open timing of Figure 5.57 are shown in Figure 5.58. For all charge motion strategies, a reduction of both the nucleation and the accumulation mode particles is obvious for the enhanced inflammation process using the HFI system.

As shown in section 2.4, accumulation mode particles of GDI engines are mostly caused by inhomogeneities in the gas phase and wall films. The wall films thereby become critical when the time for evaporation is shortened, resulting in a shorter time for the evaporation of the liquid fuel. In Figure 5.59 in the upper graphs, the start of combustion (MFB05%), centre of combustion (MFB50%) and the end of combustion

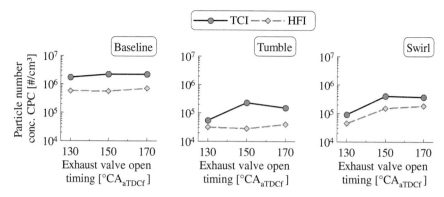

Figure 5.57: Influence of inflammation on PN emissions at 2000 rpm, 1.4 MPa IMEP, high flow injector, $p_{Inj.}$: 20 MPa, ignition timing: knock limited spark advance

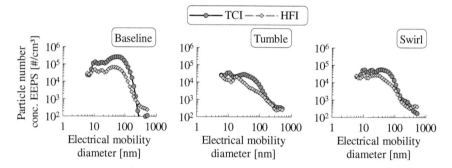

Figure 5.58: Influence of inflammation on particle size distributions at exhaust valve open timing of 130 °CA $_{aTDCf}$ at 2000 rpm, 1.4 MPa IMEP, high flow injector, $p_{Inj.}$: 20 MPa, ignition timing: knock limited spark advance

(MFB90%) are plotted versus the exhaust valve open timing for both ignition systems. It is obvious that for baseline (left hand), tumble (middle) as well as swirl configuration (right hand) the MFB50% can be advanced using the HFI system. As the engine was knock-limited for this operation point, the advanced combustion phasing results in an advance of the engines efficiency. Therefore, the fuel requirement for constant power output can be reduced, resulting in a shorter injection duration with an earlier end of injection. Additionally, the ignition timing is shifted late, caused by the advanced inflammation process, compared to the TCI system for comparable combustion phasing. These two effects increase the time for mixture preparation and thus enhance the evaporation of the wall films, which is assumed to be one reason for the reduced PN concentrations measured.

Comparing the charge motion strategies, it is necessary to mention that the engine was more prone to knocking using both charge motion configurations (tumble and swirl). Nevertheless, both the inflammation and the combustion duration (MFB05% to MFB90%) decrease with the implementation of the charge motion. Therefore, the ignition timing can be shifted later in order to further increase the time for mixture preparation. Due to the enhanced mixture formation process caused by the increased in-cylinder charge motion, the mixture shows less inhomogeneities and thus reduced PN emissions and a shorter combustion duration, as already shown in section 5.1. Furthermore, the combustion stability is advanced, as shown in the lower graphs of Figure 5.59 by the reduced values of CoV IMEP for all operating points. The implementation of large scale charge motion also reduces the cyclic variations.

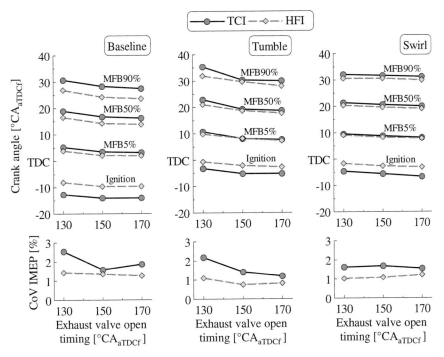

Figure 5.59: Influence of inflammation on the combustion process at 2000 rpm, 1.4 MPa IMEP, high flow injector, $p_{Inj.}$: 20 MPa, ignition timing: knock limited spark advance

Thereby the question arises, how reduced cyclic variations tend to reduce PN emissions. To answer this question, the combustion phasing (MFB50%) of the 100 cycles measured for each configuration is plotted versus the IMEP in Figure 5.60. The correlation of MFB50% and IMEP is obvious in all cases and well-established [79]. First thing to mention is the lower range of variation both in IMEP and MFB50% with the implementation of the large scale charge motion (both tumble and swirl). For all charge motion configurations, the HFI inflammation concept leads to a further decrease of variations. Thereby, especially the number of cycles with a late combustion phasing and thus indicating a low mixture homogeneity are reduced.

Additionally, the cycles with identical combustion phasing show a lower IMEP for the inflammation with the HFI system due to the increased efficiency by the earlier average combustion phasing.

Summed up, both the increased time for mixture preparation and wall film evaporation due to increased efficiency and faster inflammation as well as the reduction of the cyclic variations are assumed to cause the lowering of the PN emissions using the HFI system for inflammation.

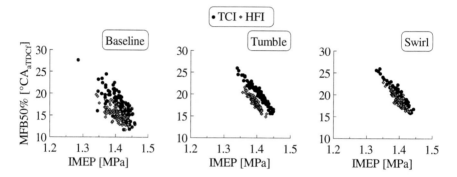

Figure 5.60: Influence of inflammation on cycle-to-cycle variations at 2000 rpm, 1.4 MPa IMEP, high flow injector, $p_{Inj.}$: 20 MPa, ignition timing: knock limited spark advance

6 Summary and conclusions

In this research work, exhaust gas analysis systems for measuring the particle number as well as the particle size distributions were combined with optical diagnostics and thermodynamic analysis to get an insight into the particle formation process in GDI engines. The single cylinder research engine used is representative for current engines that are commercially available on the market. Due to the combined usage of different investigation techniques (optical diagnostics, numerical analysis, flow field investigations, e.g.), the results can also be transferred to engines with different specifications. Additional investigations in a pressurised chamber were done to gather information on the injection process, especially on the influence of the injector hydraulic flow and the injection pressure (up to 50 MPa). Measurements at the single cylinder research engine were performed at higher engine load (WOT and boosted operation) and low engine speed (2000 rpm).

The results of this thesis help to reduce engine-out PN emissions of future GDI engine generations. As a result of the thesis, it can be stated that a "clean" combustion process, emitting PN emissions slightly above the level of ambient air, can be realised in a GDI engine at high engine load under steady state, hot engine conditions. Changes of the engine conditions, such as cold engine coolant and oil, negatively influence the PN emissions. Especially transient engine operation and related applicative settings of the engine parameters, for instance injection timing and the in-cylinder charge motion, show strong influence on the mixture formation process and PN formation. These effects are assumed to make the gasoline particle filter (GPF) necessary for upcoming engine generations to reduce PN emissions under all driving conditions. To reduce negative effects of the GPF, such as increased back-pressure, and to minimize the regeneration frequency of the GPF, the combustion system needs to be optimised to minimum engine-out PN emissions.

To realise a combustion with lowest engine out PN emissions, different influencing factors were discussed in this thesis. To provide an overview of the results shown in chapter 5, the investigations using the single cylinder research engine were used to classify the different influencing factors according to their PN reduction potential. The three groups, "STRONG INFLUENCE", "MEASURABLE INFLUENCE" and "LOW INFLUENCE" are illustrated in Figure 6.1.

One of the parameters with strongest influence on PN formation investigated in this work was the in-cylinder charge motion. Increased in-cylinder charge motion (tumble or swirl) leads to enhanced mixture formation and thus to reduced PN formation, if the level of the in-cylinder charge motion is matched to the engine. If not adjusted

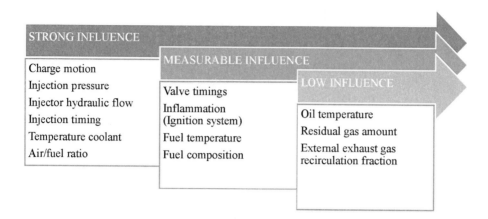

Figure 6.1: Investigated influencing factors on PN formation in GDI engines

correctly, negative effects of the implemented charge motion are possible. For instance in this work, negative effects of implemented charge motion were measured using the inlay with 70 % closed intake channel resulting in higher PN emissions and the engine being more prone to knocking. The implementation of the swirl by closing one intake runner also showed negative effects for injection timings from about -200 °CA $_{\text{aTDCf}}$ to -260 °CA $_{\text{aTDCf}}$, due to interaction of the spray with the open intake valve.

Another promising solution to reduce PN emissions is a combination of a reduced hydraulic flow of the injector with increased injection pressure. The average droplet size is reduced and the evaporation process is enhanced due to the increased surface to volume ratio with reduced hydraulic flow. By using the increased injection pressure, the fuel needed can be injected in a shorter duration with lower average droplet sizes. Due to the shorter injection time, the time for homogenisation is increased.

With reduced coolant temperature as well as with air/fuel ratio below stoichiometry, PN emissions are significantly increased. For the valve timings, the inflammation and the fuel conditions, measurable effects on PN formation were identified. Whereas for the oil temperature, the residual gas amount and the external exhaust gas recirculation fraction only minor influence on PN emissions were noted.

It is to state that the PN reduction potential of the different parameters investigated in this work were not additive and most of the effects interacted with each other. For example, a reduction of the hydraulic flow of the injector led to reduced PN emissions with the baseline configuration. With increased in-cylinder charge motion (tumble), the PN emissions were on an equal level for the two hydraulic flow variants. However, the level was lower compared to the baseline configuration (shown in Figure 5.36 on page 107).

As summary of this thesis, the following recommendations should be fulfilled to operate GDI engines with low PN emissions at high engine load:

- A short time for engine heat up is required for low PN emission engine concepts.
- The spray targeting, injection pressure and injection timing need to be optimised synchronous and adjusted to the engine operation point.
- Operation with air/fuel ratio below $\lambda = 0.9$ should be prevented.
- Large scale charge motion enhances the mixture formation process, but needs to be applied to the engine geometry and operation settings.
- The ignition system can stabilise the inflammation process and thus reduce PN emissions.
- The fuel composition influences the evaporation process, soot formation and soot oxidation. Therefore future gasoline fuel for low PN emissions should contain low amount of aromatics and a high vapour pressure.

Concerning the measurement techniques used, high-speed PIV measurements helped to quantify the flow field inside the cylinder and thus to compare the different charge motion strategies and can be used to transfer the results to different engine types. The flow field data measured in this work will be used for upcoming in-cylinder flow field simulations. These simulations will help to understand the complex interaction of charge motion and injected fuel.

The spark plug with fibre optical access is suitable to detect diffusive combustion inside the cylinder with comparatively low effort and application work. The high-speed camera measurements need more effort both in application and post processing. However, more information about the mixture formation (e.g. spray-valve interaction) and the combustion process can be gained (e.g. localisation of soot emissions).

The combination of the optical diagnostics with the thermodynamic analysis make an in-depth assessment of the mixture formation, inflammation and combustion process in recent GDI engines possible.

During this research project, some questions came up and could not be answered during the project. Such as the question about the influence of the oil composition on PN emissions. As in some operation points, the injected fuel impinges on the cylinder liner, lubricant from the liner can take part in the combustion process. Additional oil can get into the combustion chamber through the blow-by system and due to leakiness of the turbocharger bearings.

Another open question was raised about the influence of the additives in fuel and oil on PN emissions. To get an understanding of the influences of lubricant composition as well as on the additives in fuel and oil, a research proposal was submitted for an additional research project at FVV.

Bibliography

[1] ADRIAN, R. J.: Dynamic ranges of velocity and spatial resolution of particle image velocimetry. In: *Measurement Science and Technology* 8 (1997), No. 12, pp. 1393–1398.. – http://stacks.iop.org/0957-0233/8/i=12/a=003

[2] AIKAWA, K. , SAKURAI, T. , JETTER, J. J.: Development of a Predictive Model for Gasoline Vehicle Particulate Matter Emissions. In: *SAE Int. J. Fuels Lubr.* 3 (2010), No. 2. doi: 10.4271/2010-01-2115

[3] ALEIFERIS, P.G , VAN ROMUNDE, Z.R: An analysis of spray development with iso-octane, n-pentane, gasoline, ethanol and n-butanol from a multi-hole injector under hot fuel conditions. In: *Fuel* 105 (2012), pp. 143–168. doi: 10.1016/j.fuel.2012.07.044

[4] ALGER, T. , CHAUVET, T. , DIMITROVA, Z.: Synergies between High EGR Operation and GDI Systems. In: *SAE Int. J. Engines* 1 (2008), No. 1, pp. 101–114. doi: 10.4271/2008-01-0134

[5] ALGER, T. , GINGRICH, J. , KHALEK, I. A. , MANGOLD, B.: The Role of EGR in PM Emissions from Gasoline Engines. In: *SAE Int. J. Fuels Lubr.* 3 (2010), No. 1, pp. 85–98. doi: 10.4271/2010-01-0353

[6] AMER, A. , BABIKER, H. , CHANG, J. , KALGHATGI, G. , ADOMEIT, P. , BRASSAT, A. , GÜNTHER, M.: Fuel Effects on Knock in a Highly Boosted Direct Injection Spark Ignition Engine. In: *SAE Int. J. Fuels Lubr.* 5 (2012), No. 3, pp. 1048–1065. doi: 10.4271/2012-01-1634

[7] ARCOUMANIS, C. , WHITELAW, J. H.: Fluid mechanics of internal combustion engines -a review. In: *Proceedings of the Institution of Mechanical Engineers, Part C: Journal of Mechanical Engineering Science* 201 (1987), No. 1, pp. 57–74

[8] BAUMGARTEN, Carsten: *Mixture formation in internal combustion engines.* Berlin : Springer, 2006. – ISBN 978-3-540-30836-2

[9] BECK, K. W. , BERNHARDT, S. , SPICHER, U. , GEGG, T. , KÖLMEL, A. , PAA, A. , WEISSGERBER, T. , WESTECKER, M.: Ion-Current Measurement in Small Two-Stroke SI Engines. In: *SAE Technical Paper 2008-32-0037* (2008). doi: 10.4271/2008-32-0037

[10] BERNDORFER, A. , BREUER, S. , PIOCK, W. , BACHO, P. v.: Diffusion Combustion Phenomena in GDi Engines caused by Injection Process. In: *SAE Technical Paper 2013-01-0261* (2013). doi: 10.4271/2013-01-0261

[11] BERTSCH, M. , BECK, K. , ULMERICH, P. , VAN DEN HOEVEL, H. , SPICHER, U.: Comparison of the Emission Behaviour and Fuel Consumption of a Small Two-Stroke SI Chainsaw under Test-Bed- and Real In-Use Conditions. In: *SAE Technical Papers 2012-32-0070* (2012). doi: 10.4271/2012-32-0070

[12] BERTSCH, M. , BECK, K. W. , MATOUSEK, T. , SPICHER, U.: Is a High Pressure Direct Injection System a Solution to Reduce Exhaust Gas Emissions in a Small Two-Stroke Engine? In: *SAE Int. J. Engines* 6 (2013), pp. 2140–2149. doi: 10.4271/2013-32-9143

[13] BERTSCH, M , BECK, K. W. , SPICHER, U. , KÖLMEL, A. , DAWIN, U. C. , LOCHMANN, H. , SCHWEIGER, S.: Influence of the Alcohol Type and Concentration in Alcohol-Blended Fuels on the Combustion and Emission of Small Two-Stroke SI Engines. In: *SAE Technical Papers 2012-32-0038* (2012). doi: 10.4271/2012-32-0038

[14] BERTSCH, M. , KOCH, T. , VELJI, A.: Influence of charge motion and injection pressure on the particulate emission of a gasoline DI-SI engine at homogeneous, boosted operation. In: *SIA Powertrain Conference: The low CO2 spark ignition engine of the future and its hybridization* (2015)

[15] BERTSCH, M. , KOCH, T. , VELJI, A.: Untersuchungen zur Partikelemission beim Ottomotor mit Direkteinspritzung im aufgeladenen Betrieb. In: *13. FAD-Konferenz: Herausforderung Abgasnachbehandlung für Dieselmotoren* Ed. 13. Förderkreis Abgasnachbehandlungstechnologien für Dieselmotoren, 2015, pp. 85–112

[16] BERTSCH, M. , KOCH, T. , VELJI, A. , KUBACH, H.: Thermodynamic and Optical Investigations on Particle Emissions in a DISI Engine at Boosted Operation. In: *SAE Int. J. Engines* 9 (2016), No. 1, pp. 162–178. doi: 10.4271/2015-01-1888

[17] BERTSCH, M. , SCHREER, K. , DISCH, C. , BECK, K. W. , SPICHER, U.: Investigation of the Flow Velocity in the Spark Plug Gap of a Two-Stroke Gasoline Engine using Laser-Doppler-Anemometry. In: *SAE Int. J. Engines* 5 (2011), No. 4, pp. 34–41. doi: 10.4271/2011-32-0529

[18] BERTSCH, M. , WEIDENLENER, A. , DÖRNHÖFER, J. , KOCH, T. , VELJI, A.: Effect of implementing large scale charge motion, reducing hydraulic flow of the injector and increasing injection pressure on particle emissions of a GDI engine at WOT and boosted operation: submitted article. In: *International Journal of Engine Research, Special Issue: "Soot Dynamics in Internal Combustion Engines"* (2016). doi: 10.1177/1468087416670248

[19] BERTSCH, M. , WERNER, P. , KOCH, T. , HAMPE, C.: Untersuchungen der Hochfrequenz-Zündung bei unterschiedlichen Ladungsbewegungen und Ventilsteuerzeiten an einem aufgeladenen Ottomotor mit Direkteinspritzung. In: TSCHÖKE, H. (Edit): *9. Tagung Diesel- und Benzindirekteinspritzung 2014.*

Wiesbaden : Springer Vieweg, 2015, pp. 451–471. – http://link.springer.
com/book/10.1007/978-3-658-07650-4. – ISBN 978-3-658-07650-4

[20] BOCKHORN, H.: *Springer series in chemical physics.* Ed. 59: *Soot formation in combustion: Mechanisms and models.* Berlin, New York : Springer-Verlag, 1994. – ISBN 3-540-58398-X

[21] BOGARRA-MACIAS, M. , HERREROS-ARELLANO, J. M. M. , TSOLAKIS, A. , YORK, A. P. E. , MILLINGTON, P.: Reformate Exhaust Gas Recirculation (REGR) Effect on Particulate Matter (PM), Soot Oxidation and Three Way Catalyst (TWC) Performance in Gasoline Direct Injection (GDI) Engines. In: *SAE Int. J. Engines* 9 (2015). doi: 10.4271/2015-01-2019

[22] BÖHM, D. , HESSE, D. , JANDER, H. , LUERS, B. , PIETSCHER, J. , WAGNER, H. G. , WEISS, M.: The Influence of Pressure and Temperature on Soot Formation in Premixed Flames. In: *Symposium International on Combustion*, The Combustion Institute, 1988, pp. 403–412

[23] BOHNE, S. , RIXECKER, G. , BRICHZIN, V. , BECKER, M.: Hochfrequenz-Zündsystem mittels Korona-Entladung. In: *MTZ - Motortechnische Zeitschrift* 75 (2014), No. 1, pp. 50–55

[24] BURI, S.: *Untersuchungen des Potenzials von Einspritzdrücken bis 1000 bar in einem Ottomotor mit Direkteinspritzung und strahlgeführtem Brennverfahren.* Karlsruhe, Universität Karlsruhe, Dissertation, 2011

[25] BURROWS, J. , LYKOWSKI, J. , MIXELL, K.: Corona-Zündsystem für hocheffiziente Ottomotoren. In: *MTZ - Motortechnische Zeitschrift* 74 (2013), No. 6, pp. 484–487

[26] CAIRNS, A. , BLAXILL, H.: Lean Boost and External Exhaust Gas Recirculation for High Load Controlled Auto-Ignition. In: *SAE Technical Paper 2005-01-3744* (2005). doi: 10.4271/2005-01-3744

[27] CAIRNS, A. , BLAXILL, H. , IRLAM, G.: Exhaust Gas Recirculation for Improved Part and Full Load Fuel Economy in a Turbocharged Gasoline Engine. In: *SAE Technical Paper 2006-01-0047* (2006). doi: 10.4271/2006-01-0047

[28] CAIRNS, A. , FRASER, N. , BLAXILL, H.: Pre Versus Post Compressor Supply of Cooled EGR for Full Load Fuel Economy in Turbocharged Gasoline Engines. In: *SAE Technical Paper 2008-01-0425* (2008). doi: 10.4271/2008-01-0425

[29] CUDEIRO TORRUELLA, M. , SCHORR, J. , KRÜGER, C. , SAUTER, W. , PRILOP, H. , WALTNER, A. , BARGENDE, M.: High-Speed Optical Diagnostics of Soot Formation in a Spray Guided DISI Engine under Lean Stratified Operation. (2014), No. 11, pp. 100–123

[30] DAGEFÖRDE, H.: *Untersuchung innermotorischer Einflussgrößen auf die Partikelemission eines Ottomotors mit Direkteinspritzung.* Karlsruhe, Karlsruhe Institute of Technology (KIT), Dissertation, 2015. – http://digbib.ubka.uni-karlsruhe.de/volltexte/1000049421

[31] DAGEFÖRDE, H. , BERTSCH, M. , KUBACH, H. , KOCH, T.: Reduktion der Partikelemissionen bei Ottomotoren mit Direkteinspritzung. In: *MTZ - Motortechnische Zeitschrift* 76 (2015), No. 10, pp. 86–93

[32] DAGEFÖRDE, H. , KOCH, T. , BECK, K. W. , SPICHER, U.: Influence of Fuel Composition on Exhaust Emissions of a DISI Engine during Catalyst Heating Operation. In: *SAE Technical Papers 2013-01-2571* (2013). doi: 10.4271/2013-01-2571

[33] DAGEFÖRDE, H. , KOCH, T. , SPICHER, U.: *Untersuchung von Maßnahmen zur Reduktion der Partikel-Anzahlemissionen bei Otto-DI-Motoren: Abschlussbericht zu FVV-Projekt Nr. 1046.* 2014

[34] DAHL, A. , GHARIBI, A. , SWIETLICKI, E. , GUDMUNDSSON, A. , BOHGARD, M. , LJUNGMAN, A. , BLOMQVIST, G. , GUSTAFSSON, M.: Traffic-generated emissions of ultrafine particles from pavement–tire interface. In: *Atmospheric Environment* 40 (2006), No. 7, pp. 1314–1323. doi: 10.1016/j.atmosenv.2005.10.029

[35] DANZER, J. , SCHRÖDER, M. , HILL, L. , BALTES, N.: Horiba PEMS-PN: A state-of-the-art solution for RDE legislation. In: *13. FAD-Konferenz: Herausforderung Abgasnachbehandlung für Dieselmotoren.* Förderkreis Abgasnachbehandlungstechnologien für Dieselmotoren e.V., 2015, pp. 199–202. – ISSN 2199-8973

[36] DEC, J. E. , COY, E. B.: OH Radical Imaging in a DI Diesel Engine and the Structure of the Early Diffusion Flame. In: *SAE Technical Paper 960831* (1996). doi: 10.4271/960831

[37] DIERKS, U.: Accuracy of Particle Number Counting systems (PNCS) influenced by different 23nm cutpoint calibration methods. In: *17th ETH Conference on Combustion Generated Nanoparticles* (2013)

[38] DIERKSHEIDE, U. , MEYER, P. , HOVESTADT, T. , HENTSCHEL, W.: Endoscopic 2D particle image velocimetry (PIV) flow field measurements in IC engines. In: *Experiments in Fluids* 33 (2002), No. 6, pp. 794–800. doi: 10.1007/s00348-002-0499-3

[39] DISCH, C. , KUBACH, H. , SPICHER, U. , PFEIL, J. , ALTENSCHMIDT, F. , SCHAUPP, U.: Investigations of Spray-Induced Vortex Structures during Multiple Injections of a DISI Engine in Stratified Operation Using High-Speed-PIV. In: *SAE Technical Paper 2013-01-0563* (2013). doi: 10.4271/2013-01-0563

[40] DISCH, C. , PFEIL, J. , KUBACH, H. , KOCH, T. , SPICHER, U. , THIELE, O.: Experimental Investigations of a DISI Engine in Transient Operation with Regard to Particle and Gaseous Engine-out Emissions. In: *SAE Technical Papers 2015-01-1990* (2015). doi: 10.4271/2015-01-1990

[41] DRAKE, M C. , FANSLER, T D. , SOLOMON, A S. , SZEKELY, G A.: Piston Fuel Films as a Source of Smoke and Hydrocarbon Emissions from a Wall-Controlled Spark-Ignited Direct-Injection Engine. In: *SAE Technical Paper 2003-01-0547* (2003). doi: 10.4271/2003-01-0547

[42] EASTWOOD, P.: *Particulate emissions from vehicles.* Warrendale, PA and Chichester : Published on behalf of SAE International and John Wiley, 2008. – ISBN 978-0-7680-2060-1

[43] EICHLSEDER, H. , KLÜTING, M. , PIOCK, W.: *Grundlagen und technologien des ottomotors.* Wien : Springer-Verlag, 2008. – ISBN 978-3-211-25774-6

[44] EICHMEIER, J.: *Kombinierte Verbrennung brennraumintern gemischter Kraftstoffe mit unterschiedlichen Zündwilligkeiten untersucht am Beispiel von Diesel und Benzin.* Karlsruhe, Karlsruhe Institute of Technology (KIT), Dissertation, 2012

[45] ETH ZÜRICH: *Elektrische Energiesysteme: Lecture on High-Voltage Technology.* 2009. – http://www.eeh.ee.ethz.ch/uploads/tx_ethstudies/Vorlesung_Nr2_Gas_FS09_01.pdf

[46] EUROPEAN COMMISSION: *Commission welcomes Member States' agreement on robust testing of air pollution emissions by cars.* – http://europa.eu/rapid/press-release_IP-15-5945_en.htm. – Date of access: 26.01.2016

[47] EUROPEAN ENVIRONMENT AGENCY: The impact of international shipping on European air quality and climate forcing. In: *EEA Technical report* (2013), No. 4. doi: 10.2800/75763

[48] EUROPEAN ENVIRONMENT AGENCY: Air quality in Europe: 2015 report. 5 (2015). ISBN 978-92-9213-701-4. doi: 10.2800/62459

[49] FENNELL, D. , HERREROS, J. M. , TSOLAKIS, A. , XU, H. , COCKLE, K. , MILLINGTON, P.: GDI Engine Performance and Emissions with Reformed Exhaust Gas Recirculation (REGR). In: *SAE Technical Paper 2013-01-0537* (2013). doi: 10.4271/2013-01-0537

[50] FRANCQUEVILLE, L. de , PILLA, G. L.: Investigation of particle formation processes in a GDI engine in catalyst heating operation. In: *International Symposium on Combustion Diagnostics.* 10. 2012, pp. 129–137. – ISBN 9783000326691

[51] FRENKLACH, M. , WANG, H.: Detailed modeling of soot particle nucleation and growth. In: *Symposium (International) on Combustion* 23 (1991), No. 1, pp. 1559–1566. doi: 10.1016/S0082-0784(06)80426-1

[52] GADDAM, C. K. , VANDER WAL, R. L.: Physical and chemical characterization of SIDI engine particulates. In: *Combustion and Flame* 160 (2013), No. 11, pp. 2517–2528

[53] GAMMA TECHNOLOGIES INC.: *Engine Performance Application Manual: Product information, Version 7.3.* 2012

[54] GAMMA TECHNOLOGIES INC.: *Flow Theory Manual: Product information, Version 7.3.* 2012

[55] GAYDON, A. G.: *The spectroscopy of flames.* 2nd ed. London : Chapman and Hall, 1974. – ISBN 0412128705

[56] GIECHASKIEL, B.: The AVL Particle Counter: APC 489 - Experience from VPR and PNC validations. (2012). – http://www.unece.org/fileadmin/DAM/trans/doc/2011/wp29grpe/PMP-26-04e.pdf. – Date of access: 31.07.2015

[57] GIECHASKIEL, B. , CARRIERO, M. , MARTINI, G. , BERGMANN, A. , PONGRATZ, H. , JOERGL, H.: Comparison of Particle Number Measurements from the Full Dilution Tunnel, the Tailpipe and Two Partial Flow Systems. In: *SAE Technical Paper 2010-01-1299* (2010). doi: 10.4271/2010-01-1299

[58] GIECHASKIEL, B. , DILARA, P. , ANDERSSON, J.: Particle Measurement Programme (PMP) Light-Duty Inter-Laboratory Exercise: Repeatability and Reproducibility of the Particle Number Method. 42 (2008), pp. 528–543. – http://www.unece.org/fileadmin/DAM/trans/doc/2011/wp29grpe/PMP-26-04e.pdf. – Date of access: 16.11.2015

[59] GIECHASKIEL, B. , MAMAKOS, A. , ANDERSSON, J. , DILARA, P. , MARTINI, G. , SCHINDLER, W. , BERGMANN, A.: Measurement of Automotive Nonvolatile Particle Number Emissions within the European Legislative Framework: A Review. In: *Aerosol Science and Technology* 46 (2012), No. 7, pp. 719–749

[60] GIECHASKIEL, B. , MANFREDI, U. , MARTINI, G.: Engine Exhaust Solid Sub-23 nm Particles: I. Literature Survey. In: *SAE Int. J. Fuels Lubr.* 7 (2014), No. 3, pp. 950–964. doi: 10.4271/2014-01-2834

[61] GIECHASKIEL, B. , MARTINI, G.: Engine Exhaust Solid Sub-23 nm Particles: II. Feasibility Study for Particle Number Measurement Systems. In: *SAE Int. J. Fuels Lubr.* 7 (2014), No. 3, pp. 935–949. doi: 10.4271/2014-01-2832

[62] GOLLOCH, R.: *Downsizing bei Verbrennungsmotoren: Senkung des Kraftstoffverbrauchs und Steigerung des Wirkungsgrads.* 1. Berlin : Springer, 2005 (VDI-Buch). – ISBN 3-540-23883-2

[63] GRAF, J. , LAUER, T. , GERINGER, B.: Zündsysteme für hochaufgeladene Downsizingmotoren. In: *MTZ - Motortechnische Zeitschrift* 74 (2013), No. 11, pp. 898–903

[64] GRANDIN, B. , ÅNGSTRÖM, H.-E.:   Replacing Fuel Enrichment in a Turbo Charged SI Engine: Lean Burn or Cooled EGR.   In: *SAE Technical Paper 1999-01-3505* (1999). doi: 10.4271/1999-01-3505

[65] GRANDIN, B. , ÅNGSTRÖM, H.-E. , STÅLHAMMAR, P. , OLOFSSON, E.: Knock Suppression in a Turbocharged SI Engine by Using Cooled EGR. In: *SAE Technical Paper 982476* (1998). doi: 10.4271/982476

[66] GRANDIN, B. , DENBRATT, I. , BOOD, J. , BRACKMANN, C. , BENGTSSON, P.-E. , GOGAN, A. , MAUSS, F. , SUNDÉN, B.: Heat Release in the End-Gas Prior to Knock in Lean, Rich and Stoichiometric Mixtures With and Without EGR. In: *SAE Technical Paper 2002-01-0239* (2002). doi: 10.4271/2002-01-0239

[67] GRASKOW, B. R. , KITTELSON, D. B. , ABDUL-KHALEK, I. S. , AHMADI, M. R. , MORRIS, J. E.: Characterization of Exhaust Particulate Emissions from a Spark Ignition Engine. In: *SAE Technical Paper 980528* (1998). doi: 10.4271/980528

[68] GREENWOOD, S. J. , COXON, J. E. , BIDDULPH, T. , BENNETT, J.:   An Investigation to Determine the Exhaust Particulate Size Distributions for Diesel, Petrol, and Compressed Natural Gas Fuelled Vehicles. In: *SAE Technical Paper 961085* (1996). doi: 10.4271/961085

[69] HAGEMANN, M. , ROTHAMER, D.: Sensitivity Analysis of Particle Formation in a Spark-Ignition Engine during Premixed Operation. In: *Papers for the 8th U.S. National Combustion Meeting* (2013), pp. –. – http://www.che.utah.edu/~sutherland/USCI2013/PAPERS/1C07-070IC-0046.pdf. – Date of access: 01.02.2016

[70] HAK, C. , LARSSEN, S. , RANDALL, S. , GUERREIRO, C. , DENBY, B.: Traffic and Air Quality - Contribution of Traffic to Urban Air Quality in European Cities. In: *ETC/ACC Technical Paper 2009/12* (2010)

[71] HALL, D. E. , GOODFELLOW, C. L. , HEINZE, P. , RICKEARD, D. J. , NANCEKIEVILL, G. , MARTINI, G. , HEVESI, J. , RANTANEN, L. , MERINO, P. M. , MORGAN, T. D. B. , ZEMROCH, P. J.: A Study of the Size, Number and Mass Distribution of the Automotive Particulate Emissions from European Light Duty Vehicles. In: *SAE Technical Paper 982600* (1998). doi: 10.4271/982600

[72] HALL, D. E. , GOODFELLOW, H. J. , HEVESI, J. , MCARRAGHER, J. S. , MERCOGLIANO, R. , MERINO, M. P. , MORGAN, T. D. B. , NANCEKIEVILL, G. , RANTANEN, L. , RICKEARD, D. J. , TERNA, D. , ZEMROCH, P. J. , HEINZE, P.: *A study of the number, size & mass of exhaust particles emitted from european diesel and gasoline vehicles under steady-state and european driving cycle conditions: Prepared for the CONCAWE Automotive Emissions Management Group by its Special Task Force AE/STF-10.* 1998. – https://www.concawe.eu/uploads/Modules/Publications/rpt_98-51-2003-01972-01-e.pdf

[73] HAMPE, C. , BERTSCH, M. , BECK, K. W. , SPICHER, U. , BOHNE, S. , RIX-ECKER, G.: Influence of High Frequency Ignition on the Combustion and Emission Behaviour of Small Two-Stroke Spark Ignition Engines. In: *SAE Technical Paper 2013-32-9144* (2013). doi: 10.4271/2013-32-9144/

[74] HAMPE, C. , BERTSCH, M. , SPICHER, U. , BOHNE, S. , HADLER, J.: Thermo-dynamic Experiments of High-Frequency Ignition with Different Charge Motions and Valve Timings on a Supercharged Gasoline DI Engine. In: *2nd Conference on Ignition Systems for Gasoline Engines, Berlin* (2014), pp. 533–552

[75] HAMPE, C. , KUBACH, H. , SPICHER, U. , RIXECKER, G. , BOHNE, S.: Investigations of Ignition Processes Using High Frequency Ignition. In: *SAE Technical Paper 2013-01-1633* (2013). doi: 10.4271/2013-01-1633

[76] HEDGE, M. , WEBER, P. , GINGRICH, J. , ALGER, T. , KHALEK, I. A.: Effect of EGR on Particle Emissions from a GDI Engine. In: *SAE Int. J. Engines* 4 (2011), No. 1, pp. 650–666. doi: 10.4271/2011-01-0636

[77] HEISE, V. , FARAH, P. , HUSTED, H. , WOLF, E.: High Frequency Ignition System for Gasoline Direct Injection Engines. In: *SAE Technical Paper 2011-01-1223* (2011), pp. –. doi: 10.4271/2011-01-1223

[78] HENLE, A.: *Entkopplung von Gemischbildung und Verbrennung bei einem Dieselmotor*. Munich, TU München, Dissertation, 2006. – https://www.td. mw.tum.de/fileadmin/w00bso/www/Forschung/Dissertationen/henle.pdf. – Date of access: 19.01.2016

[79] HEYWOOD, J. B.: *Internal combustion engine fundamentals*. New York : McGraw-Hill, 1988 (McGraw-Hill series in mechanical engineering). – ISBN 0-07-100499-8

[80] HINDS, W. C.: *Aerosol technology: Properties, behavior, and measurement of airborne particles*. 2nd ed. New York : Wiley, 1999. – ISBN 978-0-471-19410-1

[81] HIROYASU, H. , ARAI, M.: Structures of Fuel Sprays in Diesel Engines. In: *SAE Technical Paper 900475* (1990). doi: 10.4271/900475

[82] HIROYASU, H. , ARAI, M. , TABATA, M.: Empirical Equations for the Sauter Mean Diameter of a Diesel Spray. In: *SAE Technical Paper 890464* (1989). doi: 10.4271/890464

[83] HIRSCH, A. , KIRSCHBAUM, P. , WINKLHOFER, E. , DÖHLER, A.: Sensors and evaluation methods for transient mode gasoline engine calibration. In: *International Symposium on Combustion Diagnostics* 8 (2008), pp. 271–283

[84] HOFFMANN, G. , BEFRUI, B. , BERNDORFER, A. , PIOCK, W. F. , VAR-BLE, D. L.: Fuel System Pressure Increase for Enhanced Performance of GDi Multi-Hole Injection Systems. In: *SAE Technical Paper 2014-01-1209* (2014). doi: 10.4271/2014-01-1209

[85] HOFFMEYER, H. , MONTEFRANCESCO, E. , BECK, L. , WILLAND, J. , ZIEBART, F. , MAUSS, F.: CARE – CAtalytic Reformated Exhaust Gases in Turbocharged DISI-Engines. In: *SAE Int. J. Fuels Lubr.* 2 (2009), pp. 139–148. doi: 10.4271/2009-01-0503

[86] HUEGEL, P. , KUBACH, H. , KOCH, T. , VELJI, A.: Investigations on the Heat Transfer in a Single Cylinder Research SI Engine with Gasoline Direct Injection. In: *SAE Int. J. Engines* 8 (2015), pp. 557–569. doi: 10.4271/2015-01-0782

[87] HULST, H. C. van de: *Light Scattering by Small Particles: Corrected republication of the work originally published in 1957 by Wiley.* New York : Dover Publications Inc., 1982. – ISBN 0-486-64228-3

[88] JIAO, Q. , REITZ, R. D.: Modeling of Equivalence Ratio Effects on Particulate Formation in a Spark-Ignition Engine under Premixed Conditions. In: *SAE Technical Paper 2014-01-1607* (2014). doi: 10.4271/2014-01-1607

[89] JIAO, Q. , REITZ, R. D.: Modeling soot emissions from wall films in a direct-injection spark-ignition engine. (2015), No. 8. doi: 10.1177/1468087414562008

[90] JIMENEZ, J. L. , BAHREINI, R. , COCKER, D. R. I. , ZHUANG, H. , VARUTBANGKUL, V. , FLAGAN, R. C. , SEINFELD, J. H. , O'DOWD, C. D. , HOFFMANN, T.: New particle formation from photooxidation of diiodomethane (CH_2I_2). In: *Journal of Geophysical Research* 108 (2003), No. D10, pp. 4318 ff. doi: 10.1029/2002JD002452

[91] JOINT RESEARCH CENTER: *Calibration and Modeling of PMP compliant Condensation Particle Counters: EUR 25145 EN.* – http://publications.jrc.ec.europa.eu/repository/handle/JRC67661. – Date of access: 26.01.2016

[92] JOINT RESEARCH CENTER: *Particle Measurement Programme (PMP) Light-duty Inter-laboratory Correlation Exercise (ILCE_LD) Final Report: EUR 22775 EN.* – http://publications.jrc.ec.europa.eu/repository/handle/JRC37386. – Date of access: 26.01.2016

[93] KADONO, T. , YOSHIDA, K. , SHOJI, H.: The Combustion Phenomena under Corona Discharge Application. In: *SAE Technical Paper 2002-32-1823* (2002). – http://papers.sae.org/2002-32-1823

[94] KALGHATGI, G. T.: Fuel Anti-Knock Quality - Part I. Engine Studies. In: *SAE Technical Papers 2001-01-3584* (2001). doi: 10.4271/2001-01-3584

[95] KALGHATGI, G. T.: Fuel Anti-Knock Quality- Part II. Vehicle Studies - How Relevant is Motor Octane Number (MON) in Modern Engines? In: *SAE Technical Papers 2001-01-3585* (2001). doi: 10.4271/2001-01-3585

[96] KALGHATGI, G. T. , NAKATA, K. , MOGI, K.: Octane Appetite Studies in Direct Injection Spark Ignition (DISI) Engines. In: *SAE Technical Papers 2005-01-0244* (2005). doi: 10.4271/2005-01-0244

[97] KAMOUN, H. , LAMANNA, G. , WEIGAND, B. , STEELANT, J.: *High-Speed Shadowgraphy Investigations of Superheated Liquid Jet Atomisation.* 2010

[98] KEANE, R. D. , ADRIAN, R. J.: Theory of cross-correlation analysis of PIV images. In: *Applied Scientific Research* 49 (1992), No. 3, pp. 191–215. doi: 10.1007/BF00384623

[99] KENNEDY, I. M.: Models of soot formation and oxidation. In: *Progress in Energy and Combustion Science* 23 (1997), No. 2, pp. 95–132. doi: 10.1016/S0360-1285(97)00007-5

[100] KHALEK, I. A. , BOUGHER, T. , JETTER, J. J.: Particle Emissions from a 2009 Gasoline Direct Injection Engine Using Different Commercially Available Fuels. In: *SAE Int. J. Fuels Lubr.* 3 (2010), No. 2, pp. 623–637. doi: 10.4271/2010-01-2117

[101] KINOSHITA, M. , SAITO, A. , MATSUSHITA, S. , SHIBATA, H. , NIWA, Y.: A Method for Suppressing Formation of Deposits on Fuel Injector for Direct Injection Gasoline Engine. In: *SAE Technical Paper 1999-01-3656* (1999). doi: 10.4271/1999-01-3656

[102] KITTELSON, D. B.: Engines and nanoparticles. In: *Journal of Aerosol Science* 29 (1998), No. 5-6, pp. 575–588. doi: 10.1016/S0021-8502(97)10037-4

[103] KITTELSON, D. B. , PATWARDHAN, U. , ZARLING, D. , GLADIS, D. , WATTS, W.: Issues associated with measuring nothing or almost nothing: Real-time Measurements of Metallic Ash Emissions from Engines. In: *Cambridge Particle Meeting* (2013), pp. –. – http://www.cambridgeparticlemeeting.org/sites/default/files/Presentations/2013/DKittelson%28UofMinnesota%29_2013_Issues%20asssociated%20with%20measuring%20nothing.pdf

[104] KITTELSON, D. B. , WATTS, W. F. , JOHNSON, J. P. , SCHAUER, J. J. , LAWSON, D. R.: On-road and laboratory evaluation of combustion aerosols— Part 2. In: *Journal of Aerosol Science* 37 (2006), No. 8, pp. 931–949. doi: 10.1016/j.jaerosci.2005.08.008

[105] KNEIFEL, A.: *Hochdruckeinspritzung als Möglichkeit zur Kraftstoffverbrauchs- und Abgasemissionsreduzierung bei einem Ottomotor mit strahlgeführtem Brennverfahren.* Karlsruhe, Universität Karlsruhe, Dissertation, 2008

[106] KOCH, T.: *Numerischer Beitrag zur Charakterisierung und Vorausberechnung der Gemischbildung und Verbrennung in einem direkteingespritzten, strahlgeführten Ottomotor.* Zürich, Eidgenössische Technische Hoschschule, Dissertation, 2002. – http://e-collection.library.ethz.ch/eserv/eth:26379/eth-26379-02.pdf

[107] KOHSE-HÖINGHAUS, K. , JEFFRIES, J. B.: *Applied combustion diagnostics.* New York : Taylor & Francis, 2002. – ISBN 1-56032-938-6

[108] KOLLMER, M.: *Reduktion der Partikelanzahlemission durch den Einsatz von aromatenfreiem Kraftstoff an einem aufgeladenen Ottomotor mit Direkteinspritzung.* Karlsruhe, Karlsruhe Institute of Technology, Bachelorthesis, 2015

[109] KOLMOGOROV, A. N.: The local structure of turbulence in incompressible viscous fluid for very large Reynolds numbers. In: *Proceedings of the Royal Society, London* (1991), pp. 9–13. doi: 10.1098/rspa.1991.0075

[110] KÖPPLE, F. , SEBOLDT, D. , JOCHMANN, P. , HETTINGER, A. , KUFFERATH, A. , BARGENDE, M.: Experimental Investigation of Fuel Impingement and Spray-Cooling on the Piston of a GDI Engine via Instantaneous Surface Temperature Measurements. In: *SAE Int. J. Engines* 7 (2014), No. 3, pp. 1178–1194. doi: 10.4271/2014-01-1447

[111] KÖPPLE, Fabian , JOCHMANN, Paul , HETTINGER, Alexander , KUFFERATH, Andreas , BARGENDE, Michael: A Novel CFD Approach for an Improved Prediction of Particulate Emissions in GDI Engines by Considering the Spray-Cooling on the Piston. In: *SAE Technical Papers 2015-01-0385* (2015), pp. –. doi: 10.4271/2015-01-0385

[112] KUFFERATH, A. , BERNS, S. , HAMMER, J. , BUSCH, R. , FRANK, M. , STORCH, A.: The EU 6 Challenge at GDI – Assessment of Feasible System Solutions. (2012). – http://www.övk.at/veranstaltungen_/symposien/2012/nachlese_en.pdf. – Date of access: 18.11.2015

[113] KUFFERATH, A. , KÖPPLE, F. , BARGENDE, M. , JOCHMANN, P.: Investigation of the Parameters Influencing the Spray-Wall Interaction in a GDI Engine - Prerequisite for the Prediction of Particulate Emissions by Numerical Simulation. In: *SAE Int. J. Engines* 6 (2013), pp. 911–925. doi: 10.4271/2013-01-1089

[114] KUMAR, P. , PIRJOLA, L. , KETZEL, M. , HARRISON, R. M.: Nanoparticle emissions from 11 non-vehicle exhaust sources – A review. In: *Atmospheric Environment* 67 (2013), pp. 252–277. doi: 10.1016/j.atmosenv.2012.11.011

[115] LAVOIE, G. A. , HEYWOOD, J. B. , KECK, J. C.: Experimental and Theoretical Study of Nitric Oxide Formation in Internal Combustion Engines. In: *Combustion Science and Technology* 1 (1970), No. 4, pp. 313–326

[116] LEACH, F. , STONE, R. , RICHARDSON, D.: The Influence of Fuel Properties on Particulate Number Emissions from a Direct Injection Spark Ignition Engine. In: *SAE Technical Paper 2013-01-1558* (2013). doi: 10.4271/2013-01-1558

[117] LEFEBVRE, A. H.: *Atomization and sprays.* New York : Taylor & Francis, 1989. – ISBN 0-89116-603-3

[118] LEICK, P.: *Quantitative Untersuchungen zum Einfluss von Düsengeometrie und Gasdichte auf den Primärzerfallsbereich von Dieselsprays.* Darmstadt, TU Darmstadt, Dissertation, 2008. – http://tuprints.ulb.tu-darmstadt.de/1159/3/Diss_Leick-31-10-081.pdf

[119] LEWIS, A. , AKEHURST, S. , TURNER, J. , PATEL, R. , POPPLEWELL, A.: Observations on the Measurement and Performance Impact of Catalyzed vs. Non Catalyzed EGR on a Heavily Downsized DISI Engine. In: *SAE Int. J. Engines* 7 (2014), pp. 458–467. doi: 10.4271/2014-01-1196

[120] LI, H. , GHAZI, A.K. , SOHRABI, A.: Examination of the Oil Combustion in a S.I. Hydrogen Engine. In: *SAE Technical Papers 2004-01-2914* (2004). doi: 10.4271/2004-01-2914

[121] MAIER, A. , KLAUS, U. , DREIZLER, A. , ROTTENGRUBER, H.: Fuel-Independent Particulate Emissions in an SIDI Engine. In: *SAE Int. J. Engines* 8 (2015), pp. 1334–1341. doi: 10.4271/2015-01-1081

[122] MAIER, T.: PN - Partikelanzahlmesstechnik: Die Messung von ultrafeinen Partikeln nach der heutigen europäischen LD & HD Gesetzgebung. In: *Horiba Concept*, Conference for Combustion Emissions Particulates and Testing, Dresden, 2013, pp. –

[123] MAIER, T. , KIWULL, B. , WOLF, J.-C. , WACHTMEISTER, G. , NIESSNER, R.: *Untersuchung und Bewertung verschiedener Methoden der Partikelanzahl Messung: Abschlussbericht zu FVV-Projekt Nr. 1073*. 2014

[124] MAIER, T. , WACHTMEISTER, G.: Particle Number Measurement Techniques: PMP particle number counting methodology, PMP-HD measurement system comparison. In: *14. Internationales Stuttgarter Symposium*. Stuttgart : FKFS, 2014, pp. 461–476

[125] MAIER, T. , WACHTMEISTER, G.: System comparison for PMP Particle number determination. In: *MTZ Worldwide* (2014), No. 75, pp. 50–56

[126] MARICQ, M. M. , PODSIADLIK, D. H. , BREHOB, D. D. , HAGHGOOIE, M.: Particulate Emissions from a Direct-Injection Spark-Ignition (DISI) Engine. In: *SAE Technical Papers 1999-01-1530* (1999). doi: 10.4271/1999-01-1530

[127] MARICQ, M. M. , PODSIADLIK, D. H. , CHASE, R. E.: Examination of the Size-Resolved and Transient Nature of Motor Vehicle Particle Emissions. In: *Environmental Science & Technology* 33 (1999), No. 10, pp. 1618–1626

[128] MARICQ, M. M. , PODSIADLIK, D. H. , CHASE, R. E.: Gasoline Vehicle Particle Size Distributions: Comparison of Steady State, FTP, and US06 Measurements. In: *Environmental Science & Technology* 33 (1999), No. 12, pp. 2007–2015

[129] MATOUSEK, T. , DAGEFÖRDE, H. , BERTSCH, M.: Influence of Injection Pressures up to 300 bar on Particle Emissions in a GDI-Engine. In: *17th ETH Conference on Combustion Generated Nanoparticles, Zürich* (2013)

[130] MAY, J. , BOSTEELS, D. , FAVRE, C.: An Assessment of Emissions from Light-Duty Vehicles using PEMS and Chassis Dynamometer Testing. In: *SAE Int. J. Engines* 7 (2014), pp. 1326–1335. doi: 10.4271/2014-01-1581

[131] MAYER, A.C. , ULRICH, A. , CZERWINSKI, J. , MOONEY, J. J.: Metal-Oxide Particles in Combustion Engine Exhaust. (2010). doi: 10.4271/2010-01-0792

[132] McMURRY, P. H. , WANG, X. , PARK, K. , EHARA, K.: The Relationship between Mass and Mobility for Atmospheric Particles: A New Technique for Measuring Particle Density. In: *Aerosol Science and Technology* 36 (2002), No. 2, pp. 227–238. doi: 10.1080/027868202753504083

[133] MELLING, A.: Tracer particles and seeding for particle image velocimetry. In: *Measurement Science and Technology* 8 (1997), pp. 1406–1416. – http://iopscience.iop.org/article/10.1088/0957-0233/8/12/005/pdf

[134] MIE, G.: Beiträge zur Optik trüber Medien, speziell kolloidaler Metallösungen. In: *Annalen der Physik* 330 (1908), No. 3, pp. 377–445

[135] MITTAL, V. , HEYWOOD, J. B.: *The Relevance of Fuel RON and MON to Knock Onset in Modern SI Engines.* SAE Technical Paper 2008-01-2414, 2008 doi: 10.4271/2008-01-2414

[136] MITTAL, V. , HEYWOOD, J. B.: The Shift in Relevance of Fuel RON and MON to Knock Onset in Modern SI Engines Over the Last 70 Years. In: *SAE Int. J. Engines* 2 (2009), No. 2, pp. 1–10. doi: 10.4271/2009-01-2622

[137] MITTAL, V. , REVIER, B. M. , HEYWOOD, J. B.: Phenomena that Determine Knock Onset in Spark-Ignition Engines. In: *SAE Technical Paper 2007-01-0007* (2007). doi: 10.4271/2007-01-0007

[138] MIYASHITA, K. , FUKUDA, Y. , SHIOZAKI, Y. , KONDO, K. , AIZAWA, T. , YOSHIKAWA, D. , TANAKA, D. , TERAJI, A.: TEM Analysis of Soot Particles Sampled from Gasoline Direction Injection Engine Exhaust at Different Fuel Injection Timings. In: *SAE Technical Paper 2015-01-1872* (2015). doi: 10.4271/2015-01-1872

[139] MÜLLER, S. H. R. , BÖHM, B. , GLEISSNER, M. , GRZESZIK, R. , ARNDT, S. , DREIZLER, A.: Flow field measurements in an optically accessible, direct-injection spray-guided internal combustion engine using high-speed PIV. In: *Experiments in Fluids* 48 (2010), No. 2, pp. 281–290. doi: 10.1007/s00348-009-0742-2

[140] NAGLE, J. , STRICKLAND-CONSTABLE, R. F.: Oxidation of Carbon between 1000–2000Â°C. In: *Proceedings of the Conference on Carbon* fifth (1962), pp. 154–164. doi: 10.1016/B978-0-08-009707-7.50026-1

[141] OHNESORGE, W. V.: Die Bildung von Tropfen an Düsen und die Auflsung flüssiger Strahlen. In: *Journal of Applied Mathematics and Mechanics* 16 (1936), No. 6, pp. 355–358

[142] PALAVEEV, S. , MAGAR, M. , KUBACH, H. , SCHIESSL, R. , SPICHER, U. , MAAS, U.: Premature Flame Initiation in a Turbocharged DISI Engine - Numerical and Experimental Investigations. In: *SAE Int. J. Engines* 6 (2013), No. 1, pp. 54–66. doi: 10.4271/2013-01-0252

[143] PANDEY, P. , PUNDIR, B. , PANIGRAHI, P.: Hydrogen addition to acetylene–air laminar diffusion flames: Studies on soot formation under different flow arrangements. In: *Combustion and Flame* 148 (2007), No. 4, pp. 249–262

[144] PARSONS, D. , AKEHURST, S. , BRACE, C.: The potential of catalysed exhaust gas recirculation to improve high-load operation in spark ignition engines. In: *International Journal of Engine Research* 16 (2015), No. 4, pp. 592–605. doi: 10.1177/1468087414554628

[145] PEI, Y.-Q. , QIN, J. , PAN, S.-Z.: Experimental study on the particulate matter emission characteristics for a direct-injection gasoline engine. In: *Proceedings of the Institution of Mechanical Engineers, Part D: Journal of Automobile Engineering* 228 (2014), No. 6, pp. 604–616

[146] PILCH, M. , ERDMAN, C. A.: Use of breakup time data and velocity history data to predict the maximum size of stable fragments for acceleration-induced breakup of a liquid drop. In: *International Journal of Multiphase Flow* 13 (1987), No. 6, pp. 741–757. doi: 10.1016/0301-9322(87)90063-2

[147] PIOCK, W. F. , BEFRUI, B. , BERNDORFER, A. , HOFFMANN, G.: Fuel Pressure and Charge Motion Effects on GDi Engine Particulate Emissions. In: *SAE Int. J. Engines* 8 (2015), pp. 464–473. doi: 10.4271/2015-01-0746

[148] PIOCK, W. F. , WEYAND, P. , WOLF, E. , HEISE, V.: Ignition Systems for Spray-Guided Stratified Combustion. In: *SAE Int. J. Engines* 3 (2010), No. 1, pp. 389–401. doi: 10.4271/2010-01-0598

[149] PISCHINGER, R. , KLELL, M. , SAMS, T.: *Thermodynamik der Verbrennungskraftmaschine*. 3. Wien, New York and NY : Springer, 2009 (Der Fahrzeugantrieb). – ISBN 978-3211-99276-0

[150] PÖTSCH, C. , BAUMGARTNER, L. S. , KOCH, D. , BERNHARD, F. , BEYFUSS, B. , WACHTMEISTER, G. , WICHELHAUS, D.: Optimization of the Mixture Formation for Combined Injection Strategies in High-Performance SI-Engines. In: *SAE Technical Paper 2015-24-2476* (2015). doi: 10.4271/2015-24-2476

[151] POTTEAU, S. , LUTZ, P. , LEROUX, S. , MOROZ, S. , TOMAS, E.: Cooled EGR for a Turbo SI Engine to Reduce Knocking and Fuel Consumption. In: *SAE Technical Paper 2007-01-3978* (2007). doi: 10.4271/2007-01-3978

[152] PRICE, P. , STONE, R. , OUDENIJEWEME, D. , CHEN, X.: Cold Start Particulate Emissions from a Second Generation DI Gasoline Engine. In: *SAE Technical Paper 2007-01-1931* (2007). doi: 10.4271/2007-01-1931

[153] RAFFEL, M. , WILLERT, C. E. , WERELEY, S. T. , KOMPENHANS, J.: *Particle image velocimetry: A practical guide*. 2. ed. Berlin [u.a.] : Springer, 2007. – ISBN 978-3-540-72307-3

[154] RAUBER, F.: *Experimentelle Untersuchungen zum Einfluss der Ladungsbewegung eines aufgeladenen Ottomotors mit Direkteinspritzung auf die Partikelemissionen bei hoher Motorlast*. Karlsruhe, Karlsruhe Institute of Technology, Dissertation, 2014

[155] REITZ, R. D.: *Atomization and Other Breakup Regimes of a Liquid Jet*. Princeton, USA, Princeton University, Dissertation, 1978

[156] REITZ, R. D. , BRACCO, F. V.: Mechanisms of breakup of round liquid jets. In: *Encyclopedia of fluid mechanics* 3 (1986), pp. 233–249

[157] REXEIS, M. , HAUSBERGER, S.: Trend of vehicle emission levels until 2020 – Prognosis based on current vehicle measurements and future emission legislation. In: *Atmospheric Environment* 43 (2009), No. 31, pp. 4689–4698. doi: 10.1016/j.atmosenv.2008.09.034

[158] RICARDO, H. R.: Paraffin as Fuel. In: *The Automobile Engineer* v.9 (1919), pp. 2–5

[159] RICCOBONO, F. , WEISS, M. , GIECHASKIEL, B. , BONNEL, P.: Results of the European PN-PEMS Measurement Program for the Type A Approval of Light-Duty Vehicles in Europe. In: *PEMS 2014 International Conference & Workshop* (2014). – http://www.cert.ucr.edu/events/pems2014/liveagenda/05riccobono.pdf. – Date of access: 29.01.2016

[160] RICKEARD, D. J. , BATEMAN, J. R. , KWON, Y. K. , MCAUGHEY, J. J. , DICKENS, C. J.: Exhaust Particulate Size Distribution: Vehicle and Fuel Influences in Light Duty Vehicles. In: *SAE Technical Paper 961980* (1996). doi: 10.4271/961980

[161] RIXECKER, G. , BOHNE, S. , ADOLF, M. , BECKER, M. , TRUMP, M. , BARGENDE, M.: The High Frequency Ignition System EcoFlash. In: *Advanced Ignition Systems for Gasoline Engines, Berlin* 1 (2012), pp. 65–81

[162] ROTHE, M.: *Experimentelle und numerische Analysen zum Klopfverhalten von Ottomotoren unter Volllastbedingungen*. Karlsruhe, Karlsruhe Institute of Technology, Dissertation, 2008

[163] ROTTENGRUBER, H. , TODSEN, E. C.: Potentials and limits of downsizing. In: *Knocking in Gasoline Engines* (2013), pp. 9–22

[164] SABATHIL, D. , KOENIGSTEIN, A. , SCHAFFNER, P. , FRITZSCHE, J. , DOEHLE, A.: The Influence of DISI Engine Operating Parameters on Particle Number Emissions. In: *SAE Technical Paper 2011-01-0143* (2011). doi: 10.4271/2011-01-0143

[165] SABATHIL, D. , SCHAFFNER, P. , KÖNIGSTEIN, A.: Efficient Application of Optical Measurements to reduce the Particle Emission from Direct-Injection Gasoline Engines. In: *10ᵗʰ International Symposium on Combustion Diagnostics, Baden-Baden* (2012), pp. 250–265

[166] SANKAR, S. V. , KAMEMOTO, D. Y. , BACHALO, W. D.: Sizing Large Hollow Micro-Balloons with the phase doppler interferometer. In: *Particle & Particle Systems Characterization* 10 (1993), No. 6, pp. 321–331. doi: 10.1002/ppsc.19930100606

[167] SAXENA, P. , WILLIAMS, F. A.: Testing a small detailed chemical-kinetic mechanism for the combustion of hydrogen and carbon monoxide. In: *Combustion and Flame* 145 (2006), No. 1-2, pp. 316–323

[168] SCHEFER, R. W. , WICKSALL, D. M. , AGRAWAL, A. K.: Combustion of hydrogen-enriched methane in a lean premixed swirl-stabilized burner. In: *Proceedings of the Combustion Institute* 29 (2002), pp. 843–851

[169] SCHENK, M. , FESSLER, M. , ROTTENGRUBER, H. , FISCHER, H.: Comparison of the thermodynamic potential of alternative ignition systems for SI-engines. In: *10ᵗʰ International Symposium on Combustion Diagnostics, Baden-Baden* 10 (2012), pp. 138–157

[170] SCHILLING, A.: *Motor Oils And Engine Lubrication.* Scientific Publications, 1968. – ISBN 978-0900645075

[171] SCHNEIDER, B.: *Experimentelle Untersuchungen zur Spraystruktur in Transienten, verdampfenden und nicht verdampfenden Brennstoffstrahlen und Hochdruck.* Zürich, ETH Zürich, Dissertation, 2003

[172] SCHULZ, F. , SCHMIDT, J. , KUFFERATH, A. , SAMENFINK, W.: Gasoline Wall Films and Spray/Wall Interaction Analyzed by Infrared Thermography. In: *SAE Int. J. Engines* 7 (2014), No. 3, pp. 1165–1177. doi: 10.4271/2014-01-1446

[173] SCHUMANN, F.: *Experimentelle Grundlagenuntersuchungen zum Katalysatorheizbetrieb mit strahlgeführter Benzin-Direkteinspritzung und Einspritzdrücken bis 800 bar.* Karlsruhe, Karlsruhe Institute of Technology, Dissertation, 2014

[174] SCHUMANN, F. , SARIKOC, F. , BURI, S. , KUBACH, H. , SPICHER, U.: Potential of spray-guided gasoline direct injection for reduction of fuel consumption and simultaneous compliance with stricter emissions regulations. (2013), No. 1. doi: 10.1177/1468087412451695

[175] SCHWENGER, C. , SPICHER, U. , VELJI, A.: A new approach of correlation between optically-measured soot concentrations in the combustion chamber and soot emissions in the exhaust. In: *8ᵗʰ International Symposium on Combustion Diagnostics, Baden-Baden* 8 (2008), pp. 71–88

[176] SHIRAISHI, T. , KAKUHO, A. , URUSHIHARA, T. , CATHEY, C. , TANG, T. , GUNDERSEN, M.: A Study of Volumetric Ignition Using High-Speed Plasma for Improving Lean Combustion Performance in Internal Combustion Engines. In: *SAE Int. J. Engines* 1 (2008), pp. 399–408. doi: 10.4271/2008-01-0466

[177] STANSFIELD, P. , WIGLEY, G. , JUSTHAM, T. , CATTO, J. , PITCHER, G.: PIV analysis of in-cylinder flow structures over a range of realistic engine speeds. In: *Experiments in Fluids* 43 (2007), No. 1, pp. 135–146. doi: 10.1007/s00348-007-0335-x

[178] STARIKOVSKAIA, S. M.: Plasma assisted ignition and combustion. In: *Journal of Physics D: Applied Physics* 39 (2006), No. 16, pp. R265–R299

[179] STARIKOVSKAIA, S. M. , STARIKOVSKII, A. Y.: Plasma-Assisted Ignition and Combustion. In: *Handbook of combustion* 5 (2010), No. 4, pp. 71–93

[180] STEIMLE, F. , KULZER, A. , RICHTER, H. , SCHWARZENTHAL, D. , ROMBERG, C.: Systematic Analysis and Particle Emission Reduction of Homogeneous Direct Injection SI Engines. In: *SAE Technical Paper 2013-01-0248* (2013). doi: 10.4271/2013-01-0248

[181] STIESCH, G.: *Modeling engine spray and combustion processes*. Berlin and New York : Springer, 2003. – ISBN 3-540-00682-6

[182] STONE, R. , ZHAO, H. , ZHOU, L.: Analysis of Combustion and Particulate Emissions when Hydrogen is Aspirated into a Gasoline Direct Injection Engine. In: *SAE Technical Paper 2010-01-0580* (2010). doi: 10.4271/2010-01-0580

[183] STUMPF, Markus , VELJI, Amin , SPICHER, Ulrich , JUNGFLEISCH, Beate , SUNTZ, Rainer , BOCKHORN, Henning: Investigations on Soot Emission Behavior of A Common-Rail Diesel Engine during Steady and Non-Steady Operating Conditions by Means of Several Measuring Techniques. In: *SAE Technical Paper 2005-01-2154* (2005). doi: 10.4271/2005-01-2154

[184] TAN, C. , XU, H. , MA, H. , GHAFOURIAN, A.: Investigation of VVT and spark timing on combustion and particle emission from a GDI Engine during transient operation. In: *SAE Technical Paper 2014-01-1370* (2014). doi: 10.4271/2014-01-1370

[185] TANG, Q. , LIU, J. , ZHAN, Z. , HU, T.: Influences on Combustion Characteristics and Performances of EGR vs. Lean Burn in a Gasoline Engine. In: *SAE Technical Paper 2013-01-1125* (2013). doi: 10.4271/2013-01-1125

[186] TANG, X. , KABAT, D. M. , NATKIN, R. J. , STOCKHAUSEN, W. F. , HEFFEL, J.: Ford P2000 Hydrogen Engine Dynamometer Development. In: *SAE Technical Paper 2002-01-0242* (2002). doi: 10.4271/2002-01-0242

[187] TOPINKA, J. A. , GERTY, M. D. , HEYWOOD, J. B. , KECK, J. C.: Knock Behavior of a Lean-Burn, H2 and CO Enhanced, SI Gasoline Engine Concept. In: *SAE Technical Paper 2004-01-0975* (2004). doi: 10.4271/2004-01-0975

[188] TOWERS, C. E. , TOWERS, D. P.: Cyclic variability measurements of in-cylinder engine flows using high-speed particle image velocimetry. In: *Measurement Science and Technology* 15 (2004), No. 9, pp. 1917. – http: //stacks.iop.org/0957-0233/15/i=9/a=032

[189] TSI: Engine Exhaust Particle Sizer Spectrometer Model 3090: Technical Specifications. (2014). – http://www.tsi.com/ engine-exhaust-particle-sizer-spectrometer-3090/. – Date of access: 06.08.2015

[190] TURNER, J. W. G. , PEARSON, R. J. , CURTIS, R. , HOLLAND, B.: Effects of Cooled EGR Routing on a Second-Generation DISI Turbocharged Engine Employing an Integrated Exhaust Manifold. In: *SAE Technical Paper 2009-01-1487* (2009). doi: 10.4271/2009-01-1487

[191] UNITED NATIONS ECONOMOIC COMMISSION FOR EUROPE (UN-ECE): *Worldwide harmonized Light vehicles Test Procedure (WLTP): UN/ECE/WP.29/GRPE/WLTP-IG.* – https://www2.unece.org/ wiki/download/attachments/29229449/GRPE-72-02%20-%20Technical% 20Report%20Phase%201b%20-%20Final%20version.pdf?api=v2. – Date of access: 26.01.2016

[192] UNITED NATIONS ECONOMOIC COMMISSION FOR EUROPE (UNECE): *Uniform provisions concerning the approval of vehicles with regard to the emission of pollutants according to engine fuel requirements.* 2008. – http://eur-lex.europa.eu/LexUriServ/LexUriServ.do?uri=OJ:L:2008: 119:0001:0181:EN:PDF

[193] VELJI, A. , YEOM, K. , WAGNER, U. , SPICHER, U. , ROSSBACH, M. , SUNTZ, R. , BOCKHORN, H.: Investigations of the Formation and Oxidation of Soot Inside a Direct Injection Spark Ignition Engine Using Advanced Laser-Techniques. In: *SAE Technical Paper 2010-01-0352* (2010). doi: 10.4271/2010-01-0352

[194] VÍTEK, O. , MACEK, J. , POLÁŠEK, M. , SCHMERBECK, S. , KAMMERDIENER, T.: Comparison of Different EGR Solutions. In: *SAE Technical Paper 2008-01-0206* (2008). doi: 10.4271/2008-01-0206

[195] VLACHOS, T. G. , BONNEL, P. , WEISS, M. , GIECHASKIEL, B. , RICCOBONO, F.: New Euro 6 Legislation on RDE Overview and Technical Challenges. In: *European Commission - Joint Research Centre (JRC); AECC RDE Seminar JRC* (2015). – http://www.aecc.eu/content/RDE_seminar/04%20-% 20AECC%20RDE%20Seminar%20JRC%20New%20Euro%206%20Legislation%20on% 20RDE%20Overview%20and%20Technical%20Challenges.pdf. – Date of access: 29.01.2016

[196] WANKER, R. , VITALE, G.: RDE - eine Herausforderung für die Messtechnik und den Entwicklungsprozess. In: *13. FAD-Konferenz: Herausforderung Abgasnachbehandlung für Dieselmotoren, Dresden.* Förderkreis Abgasnachbehandlungstechnologien für Dieselmotoren e.V., 2015, pp. 189–198. – ISSN 2199-8973

[197] WARNATZ, J. , MAAS, U. , DIBBLE, R. W.: *Combustion: Physical and chemical fundamentals, modeling and simulation, experiments, pollutant formation.* 4. ed. Berlin and Heidelberg [u.a.] : Springer, 2006. – ISBN 3-540-25992-9

[198] WEBER, D. , LEICK, P.: Structure and Velocity Field of Individual Plumes of Flashing Gasoline Direct Injection Sprays. In: *26th Annual Conference on Liquid Atomization and Spray Systems, Bremen* (2014)

[199] WERNER, P.: *Maßnahmen zur Reduktion der Partikelanzahlemissionen an einem Ottomotor mit Direkteinspritzung bei hoher Last.* Karlsruhe, Karlsruhe Institute of Technology, Masterthesis, 2014

[200] WESTERWEEL, J.: Fundamentals of digital particle image velocimetry. In: *Measurement Science and Technology* 8 (1997), No. 12, pp. 1379–1392. doi: 10.1088/0957-0233/8/12/002

[201] WESTERWEEL, J.: Theoretical analysis of the measurement precision in particle image velocimetry. In: *Experiments in Fluids* 29 (2000), No. 7, pp. S003–S012. doi: 10.1007/s003480070002

[202] WETZEL, J. , HENN, M. , GOTTHARDT, M. , ROTTENGRUBER, H.: Experimental Investigation of the Primary Spray Development of GDI Injectors for Different Nozzle Geometries. In: *SAE Technical Paper 2015-01-0911* (2015). doi: 10.4271/2015-01-0911

[203] WIESE, W. , KUFFERATH, A. , STORCH, A. , ROGLER, P.: Requirements for multi-hole injectors to meet future emission standards in direct-injection gasoline engines. In: *2nd International Engine Congress* Ed. ISBN 978-3-658-08861-3. 2015, pp. 63–80

[204] WILLERT, C. , STASICKI, B. , KLINNER, J. , MOESSNER, S.: Pulsed operation of high-power light emitting diodes for imaging flow velocimetry. In: *Measurement Science and Technology* 21 (2010), No. 7, pp. 75402. doi: 10.1088/0957-0233/21/7/075402

[205] WINKLER, M. , GRIMM, J. , LENGA, H. , HARTMANN, R. , HYOUK MIN, B. ; AACHEN COLLOQUIUM AUTOMOBILE AND ENGINE TECHNOLOGY (Edit): *Low Pressure EGR for Downsized Gasoline Engines.* 2014

[206] WOLF, T. , SCHENK, M. , SCHRÖTER, M. , ZELLINGER, F. , KLAUS, B. , PFEIFFER, D. , FISCHER, H.: RF Corona Ignition vs. Spark Ignition: A Comparison for Varying Thermodynamic Conditions and Combustion Strategies of Modern Turbocharged Gasoline Engines. In: *2nd Conference on Ignition Systems for Gasoline Engines, Berlin* (2014), pp. 503–532

[207] Xu, K. , Xie, H. , Wan, M. , Chen, T. , Zhao, H.: Effect of Valve Timing and Residual Gas Dilution on Flame Development Characteristics in a Spark Ignition Engine. In: *SAE Int. J. Engines* 7 (2014), No. 1, pp. 488–499. doi: 10.4271/2014-01-1205

Lebenslauf

Persönliche Daten

Name:	Markus Bertsch
Geburtsdatum:	26.09.1985
Geburtsort:	Karlsruhe
Eltern:	Franz Josef Bertsch
	Susanne Michaela Bertsch, geb. Kaspar

Schulbildung

1993 - 1997	Friedrichschule Durmersheim
1997 - 2001	Wilhelm-Hausenstein-Gymnasium Durmersheim
2001 - 2005	Otto-Hahn-Gymnasium Karlsruhe

Studium

2005 - 2011	Maschinenbaustudium, Universität Karlsruhe (TH)
	Vertiefungsrichtung: Allgemeiner Maschinenbau
	Abschluss: Diplom-Ingenieur

Berufliche Laufbahn

2011 - 2013	Motor, Optik, Thermodynamik Forschungs- und Entwicklungsgesellschaft mbH (MOT GmbH), Karlsruhe
2013 - 2016	Wissenschaftlicher Mitarbeiter am Institut für Kolbenmaschinen, Karlsruher Institut für Technologie (KIT)
seit Mai 2016	Robert Bosch GmbH, Schwieberdingen

Logos Verlag Berlin

ISBN 978-3-8325-4403-4

ISSN 1615-2980

ISBN 978-3-8325-4403-4

9 783832 544034 >